SpringerBriefs in Climate Studies

SpringerBriefs in Climate Studies present concise summaries of cutting-edge research and practical applications. The series focuses on interdisciplinary aspects of Climate Science, including regional climate, climate monitoring and modeling, palaeoclimatology, as well as vulnerability, mitigation and adaptation to climate change. Featuring compact volumes of 50 to 125 pages (approx. 20,000–70,000 words), the series covers a range of content from professional to academic such as: a timely reports of state-of-the art analytical techniques, literature reviews, in-depth case studies, bridges between new research results, snapshots of hot and/ or emerging topics.

Author Benefits: SpringerBriefs in Climate Studies allow authors to present their ideas and readers to absorb them with minimal time investment. Books in this series will be published as part of Springer's eBook collection, with millions of users worldwide. In addition, Briefs will be available for individual print and electronic purchase. SpringerBriefs books are characterized by fast, global electronic dissemination and standard publishing contracts. Books in the program will benefit from easy-to-use manuscript preparation and formatting guidelines, and expedited production schedules. Both solicited and unsolicited manuscripts are considered for publication in this series. Projects will be submitted to editorial review by editorial advisory boards and/or publishing editors. For a proposal document, please contact the Publisher.

K. Naveena • Ramiz Tasiya • Shilpesh Rana

Spatio-temporal Trend Analysis of Rainfall using R Software and ArcGIS

A Case Study of an Agro-climatic Zone-1 of Gujarat, India

 Springer

K. Naveena
Centre for Water Resources
Development and Management
Land and Water Management
Research Group
Kozhikode, Kerala, India

Ramiz Tasiya
Civil Engineering Consultant
Godhra, Gujarat, India

Shilpesh Rana
Civil Engineering Department
Faculty of Technology & Engineering,
The Maharaja Sayajirao University
of Baroda
Vadodara, Gujarat, India

ISSN 2213-784X   ISSN 2213-7858   (electronic)
SpringerBriefs in Climate Studies
ISBN 978-3-031-48258-8   ISBN 978-3-031-48259-5   (eBook)
https://doi.org/10.1007/978-3-031-48259-5

This Springer imprint is published by the registered company Springer Nature Switzerland AG
The registered company address is: Gewerbestrasse 11, 6330 Cham, Switzerland

Paper in this product is recyclable.

Acknowledgments

First and foremost, we would like to thank **God** for the blessings that He has bestowed upon us in all our endeavors.

We would then like to thank **Dr. Manoj Samuels**, Director, Centre of Water Resources Development and Management (CWRDM), Kerala, for permitting us to carry out our research work collaboratively.

We would also like to thank the Head of the Civil Engineering Department and the Dean of the Faculty of Technology and Engineering of M.S. University of Baroda for their kind permission to utilize the department infrastructural facilities such as library and computer lab during the research work.

Finally, we would like to thank State Water Data Centre (**SWDC**) in Gujarat and Indian Meteorological Department (**IMD**) in Pune for providing the required data as well as **USGS** and **NASA** for providing the required online data through the internet.

Contents

About the Authors

K. Naveena is a highly accomplished scientist at the Centre for Water Resource Development and Management (CWRDM). With a Ph.D. and an M.Sc. in Agricultural Statistics, he has received prestigious awards such as the Young Achiever Award from the Society for Advancement of Human and Nature, the SP Dhall Distinguished Publication Award in Statistics, and the Young Scientist Award from the ASTHA Foundation. His research interests cover machine learning modeling, forecasting modeling, and design of experiment. He has contributed significantly to the scientific knowledge in these areas through over 20 publications in renowned journals.

Ramiz Tasiya is presently working as a civil engineering consultant. He completed M.E. Civil-Hydraulic structures from Civil Engineering Department, Faculty of Technology and Engineering, the M.S. University of Baroda, Vadodara. He has published two research papers in good quality Scopus indexed journals.

Shilpesh Rana is presently working as Assistant Professor in Civil Engineering Department at the Faculty of Technology and Engineering, the M.S. University of Baroda, and he completed his PhD from SVNIT Surat. His areas of research are water resources engineering and climate change. He has published more than 11 research papers in good quality reputed Scopus indexed journals. He has guided seven ME students for the dissertation.

Acronyms

ArcGIS	Aeronautical Reconnaissance Coverage Geographic Information System
DEM	Digital Elevation Model
GCS	Geographic Coordinate System
IDW	Inverse Distance Weighting
IGBP	International Geosphere-Biosphere Programme
IPCC	Intergovernmental Panel on Climate Change
ITA	Innovative Trend Analysis
LR	Linear Regression
LU/LC	Land Use/Land Cover
MK	Mann-Kendall
PCS	Projected Coordinate System
SRTM	Shuttle Radar Topography Mission
SWDC	State Water Data Centre
USGS	United States Geological Survey
UTM	Universal Transverse Mercator

Chapter 1
Introduction

1.1 Background

During changing climatic scenarios, the precise identification of rainfall movements is very essential for planning and implementing various activities, including agricultural practices, cropping systems, drought, and flood management (Hassan et al. 2020). Seasonal rainfall variation is the most unique impact of climate change, especially in the monsoon period (June to September) in India. Long-term studies conducted on a yearly basis are therefore essential for taking early preventive measures (Mehta et al. 2019).

A humid climate has been observed in the southern districts and a dry climate in the northern districts. Based on rainfall distribution, the climate in Gujarat can be portioned into three seasons, the winter season runs from November to February, the summer from March to May, and the southwest monsoon season from June to September (Vaidya et al. 2012).

The monsoon season accounts for 95% of the total rainfall in Gujarat, with July (40%) and August (28%) contributing the most. Trend analysis of the previous data is a common method used to assess climate change. Extensive literature is available on the application of trend analysis techniques to the past datasets of climatic parameters (Mun-Ju Shin et al. 2013). Long-term rainfall movement analysis by considering the similarity of rainfall distribution over a cross-section is needed for better water resource management activities. A review of these studies shows that various parametric, nonparametric, and graphical methods have been employed to analyze trends in meteorological time series.

Heavy rainfall within a short span of the period has been observed in coastal Maharashtra, Gujarat, and Kerala in recent years (Saha et al. 2018).

Among the different states in India, Gujarat is also facing frequent drought conditions due to uncertainty in rainfall distribution (Koyel and Lunagari 2020), where the average annual rainfall varies from 330 mm to 1520 mm (Dhorde et al. 2016).

K. Naveena et al., *Spatio-temporal Trend Analysis of Rainfall using R Software and ArcGIS*, SpringerBriefs in Climate Studies,
https://doi.org/10.1007/978-3-031-48259-5_1

For an agricultural country like India, not only the magnitude but also the timely occurrence of the rainfall is equally important. Agricultural activities in India are at high risk due to climate change and due to extreme climatic events like drought as 67% of Indian agricultural land is directly dependent on rainfall. Furthermore, the water supply for these activities through irrigation systems and groundwater is also dependent on rainfall.

1.2 Impacts of Climate Change

As per the IPCC (Intergovernmental Panel on Climate Change), the following impacts of climate change have been observed:

- Global increase in temperature.
- An average rise in sea level of 1.8 mm/year between 1961 and 2003 and 3.1 mm/year between 1993 and 2003.
- Increased cyclones in the North Atlantic region.
- Changes in the rainfall pattern.
- Increase in extreme rainfall events such as droughts and floods.
- Increase in temperature- and heat-related diseases.

The mean temperature in India has been increasing at a rate of 0.51 °C/year in the last 100 years. The most sensitive states to the rise in seawater level are Tamil Nadu, Gujarat, Andhra Pradesh, Orissa, West Bengal, and Pondicherry. India has also been observing an increase in extreme rainfall events between 1951 and 2007.
Source (https://ccd.gujarat.gov.in/observed-changes.htm).

1.3 Challenges in Agriculture Due to Climate Change

The possible challenges that the agricultural sector of India is likely to face due to climate change are:

1. The challenge of water availability due to shifts and changes in the rainfall pattern and increased crop water requirements due to increased temperature.
2. The rise in the occurrences of extreme rainfall events such as floods and droughts that have severe impacts on the agricultural sector.
3. As per the report of IPCC, India is likely to face decreased seasonal precipitation and increased extreme precipitation events (floods and droughts) during the monsoon season.

To sustain and improve the production of crops in semiarid regions, we need to utilize the knowledge gained from climate variability and innovative cropping methods (Source: https://ccd.gujarat.gov.in/definitions-climat.html).

1.4   Causes of Climate Change

Climate changes and fluctuations in the climate have been observed since 4.6 billion years ago. However, the changes observed in the last century due to changes in greenhouse gases are very rapid. The major causes for these changes in climate can be divided into two groups:

1. Anthropogenic: This consists of fossil fuel use, industrialization, deforestation, waste and wastewater, etc.
2. Natural: This consists of forest fires, continental drifts, volcanic eruptions, earth tilts, etc.

1.5   Agro-Climatic Zone

Agro-climatic zoning is a part of the regionalization process. The reason to classify the regions based on their agro-climatic characteristics is to outline comparable resources, for producing and transferring agro-technology to meet the country's requirements of food, fodder, and fiber. Most early attempts at regionalization were on the basis of broad natural regions, existing cropping patterns, as well as a broad framework of climatic variations at a macro scale. The state studied in this research, i.e., Gujarat, has been divided into eight agro-climatic zones as shown in Fig. 1.1. The region falling under each zone is shown in Table 1.1.

1.6   Aim of the Study

The aim of this research work is to detect the changes in the rainfall trend by analyzing the long-term rainfall data. Rainfall is the key parameter affecting crop productivity. The semiarid region selected as the study area is a leading producer of rice, maize, jowar, and groundnuts across Gujarat as per the data obtained from the Directorate of Agriculture. This research will be useful in understanding the rainfall changes in the past and the impacts on the cropping patterns.

1.7   Objectives of the Study

The main objective of the research work is to detect the changes in the monthly and SW monsoon rainfall over the semiarid region of Gujarat.

Fig. 1.1 Agro-climatic zones of Gujarat. (Source: https://www.mapsofindia.com/maps/gujarat/gujarat-agro-climate-zone-map.html)

Table 1.1 A list of the agro-climatic zones of Gujarat

Southern heavy rainfall area and hilly area
Semiarid to dry sub-humid climate
South Gujarat
Semiarid to dry sub-humid climate
Middle Gujarat
Semiarid climate
North Gujarat
Arid to semiarid climate
Northwest Gujarat
Arid to semiarid climate
North Saurashtra
Dry sub-humid climate
South Saurashtra
Dry sub-humid climate
Bhal and coastal area

Source: Agriculture & Co-operation Department, Govt. of Gujarat

To achieve the main objective of the research, we divided it into various sub-objectives as seen below:

1. To categorize the stations into "clusters" or groups to identify the similarity in rainfall patterns.
2. Analysis of SW monsoon and monthly rainfall data using Mann-Kendall test and Sen's slope estimator test.
3. Spatial trend analysis of rainfall and rainfall trend using the inverse distance weighting (IDW) method in ArcGIS 10.3.
4. To prepare the digital elevation model of the study area using ArcGIS 10.3.
5. To prepare land use/land cover maps at decadal intervals for 1985, 1995, and 2005 for the semiarid region using ArcGIS 10.3.
6. To relate the land use/land cover map with the results of trend analysis and finding the possible impacts of climate change.

1.8   Limitations

1. The results and the conclusion are based on the analysis carried out on the data of 30 rain gauge stations. If a greater number of rain gauge stations are used, then the conclusions can be further improved.
2. The analysis is done with data available after 1970. If the time span is changed, then the results and conclusion may change.

References

Hassan M, Noreen Z, Rashid A (2020) Regional frequency analysis of annual daily rainfall maxima in Skane. Sweden Inter J Climatol:1–14. https://doi.org/10.1002/joc.7074

Koyel S, Lunagari M (2020) Association between drought and agricultural productivity using remote sensing data: a case study of Gujarat state of India. J Water Climate Change 11(S1):189–202. https://doi.org/10.2166/wcc.2020.157

Mehta L, Srivastava S, Adam HN, Alankar SB, Ghosh U, Kumar VV (2019) Climate change and uncertainty from 'above' and 'below': perspectives from India. J Regional Environ Change 19:1533–1547

Saha S, Chakraborty D, Paul RK et al (2018) Disparity in rainfall trend and patterns among different regions: analysis of 158 years' time series of rainfall dataset across India. Theor Appl Climatol 134:381–395. https://doi.org/10.1007/s00704-017-2280-9

Shin M-J, Joseph HA, Guillaume FW, Croke AJ, Jakema (2013) Addressing ten questions about conceptual rainfall–runoff models with global sensitivity analyses in R. J Hydrol 503:135–152. https://doi.org/10.1016/j.jhydrol.2013.08.047

Vaidya VB, Suvarn D, Kulshreshtha MS (2012) Evaluation of frequency analysis of distinctive rainfall intensity for various stations of Gujarat. IJISET - Inter J Innovat Sci Eng Technol 7(12):2348–7968

Chapter 2
Literature Survey

The research work aims to analyze the long-term spatiotemporal rainfall trend over the semiarid region of Gujarat. The agro-climatic regions are classified by relating the cropping patterns with the climatic variability of the region. In this research, an attempt is made to understand the variation of long-term rainfall in the semiarid region using cluster analysis. It is seen from the rainfall trend analysis that the rainfall trend correlates with the land use/land cover patterns. The spatial variation of the mean rainfall and rainfall trend is also analyzed to understand the patterns of rainfall within the study area.

During changing climatic scenarios, the precise identification of rainfall movements is very essential for planning and implementing various activities, including agricultural practices, cropping systems, drought, and flood management (Hassan et al. 2020).

Tasiya et al. (2023a, b) examined the spatiotemporal variation of rainfall in the semiarid region of Gujarat using the modified Mann-Kendall (MK) test and Inverse Distance Weighting (IDW) method.

Seasonal rainfall variation is the most unique impact of climate change, especially in the monsoon period (June to September) in India. Long-term studies conducted every year are therefore essential for taking early preventive measures (Lyla Mehta et al. 2019).

The southwest monsoon season that lasts from June to the end of September contributes to 90% of the annual rainfall in the state of Gujarat (Bandyopadhyay et al. 2016).

Patel et al. (2021) carried out a spatial and temporal trend analysis of seasonal and annual rainfall in the Bhogavo River watersheds in the Sabarmati lower basin of India. The analysis was done with the rainfall data from 11 rain gauge stations using the MK test, Sen's slope estimator test, innovative trend analysis method, and linear regression. To detect the spatial variability of rainfall, the IDW method was used in ArcGIS 10.3.

© The Author(s), under exclusive license to Springer Nature Switzerland AG 2023
K. Naveena et al., *Spatio-temporal Trend Analysis of Rainfall using R Software and ArcGIS*, SpringerBriefs in Climate Studies,
https://doi.org/10.1007/978-3-031-48259-5_2

Jeneiova et al. (2014) analyzed the rising awareness and concerns about flood risks in the Slovak Republic using the MK test.

Tasiya et al. (2023a, b) examined the temporal variability of rainfall using the innovative trend analysis method. The rainfall trend analysis using long-term rainfall data is done in the agro-climatic region of Gujarat. The possible change point in the semiarid region was analyzed using the Standard Homogeneity Point (SNHP) test.

Gocic and Trajkovic (2012) analyzed the seasonal and annual rainfall variability of seven climatic parameters for 12 weather stations. The MK test with Sen's slope estimator test was used to spot the trend of rainfall.

Suryavanshi et al. (2014) investigated climatic variables for the Betwa basin located in Central India. Climatic variables such as precipitation, potential evapotranspiration, and temperature were analyzed. The results of this study can be used for the preparation of development and management plans for the Betwa basin.

Hao and Hui (2016) investigated the trend of rainfall for 14 rain gauge stations in the Shaanxi region of China. Simple linear regression, MK test, and innovative trend analysis methods were used for the analysis. While comparing the various methods, it is observed that innovative trend analysis is better as it gives graphical representation with an indication of subtrends.

The spatial variability of soil salinity was analyzed using the IDW method and the cokriging method. This study aimed at finding the most appropriate method to model the spatial variability of soil salinity. From the analysis, it was found that IDW interpolation was more accurate than the cokriging method when using surface salt content to predict the salt content in deep layers.

References

Bandyopadhyay N, Bhuiya C, Saha AK (2016) Heat waves, temperature extremes and their impacts on monsoon rainfall and meteorological drought in Gujarat, India. Nat Hazards 82:367–388. https://doi.org/10.1007/s11069-016-2205-4

Gocic M, Trajkovic S (2012) Analysis of changes in meteorological variables using Mann-Kendall and Sen's slope estimator statistical tests in Serbia. Glob Planet Chang 100:172–182. https://doi.org/10.1016/j.gloplacha.2012.10.014

Hao W, Hui Q (2016) Innovative trend analysis of annual and seasonal rainfall and extreme values in Shaanxi, China, since the 1950s. Int J Climatol. https://doi.org/10.1002/joc.4866

Hassan M, Noreen Z, Rashid A (2020) Regional frequency analysis of annual daily rainfall maxima in Skane. Sweden Inter J Climatol:1–14. https://doi.org/10.1002/joc.7074

Jeneiova K, Kohnova S, Miroslav Sabo Detecting (2014) Trends in the annual maximum discharges in the Vah River basin. Slovakia Acta Silvatica et Lignaria Hungarica:133–144. https://doi.org/10.2478/aslh-2014-0010

Mehta L, Srivastava S, Adam HN, Alankar SB, Ghosh U, Kumar VV (2019) Climate change and uncertainty from 'above' and 'below': perspectives from India. J Regional Environ Change 19:1533–1547

Patel PS, Rana SC, Josh GS (2021) Temporal and spatial trend analysis of rainfall on Bhogavo River watersheds in Sabarmati lower basin of Gujarat, India. Acta Geophys 69:353–364. https://doi.org/10.1007/s11600-020-00520-2

Suryavanshi S, Pandey A, Chaube UC et al (2014) Long-term historic changes in climatic variables of Betwa Basin, India. Theor Appl Climatol 117:403–418. https://doi.org/10.1007/s00704-013-1013-y

Tasiya RF, Naveena K, Rana SC (2023a) Rainfall trend analysis in Gujarat's semi-arid zone: a modified approach with auto-correlation consideration. Inter J Hydrol Sci Technol. https://doi.org/10.1504/IJHST.2023.10056997

Tasiya RF, Rana SC, Naveena K (2023b) Change point and trend analysis of rainfall for the Semi-Arid zone of Gujarat state. Water Energy Inter 65(11):6–14

Chapter 3
Study Area and Data Collection

3.1 Study Area

The semiarid region of Gujarat has been selected as the study area for the research work. The study area consists of a major portion of Middle Gujarat and part of the Charotar region. The study area covers Anand, Kheda, Dahod, Chhota Udepur, Panchmahals and Vadodara districts. The geographical area covered by the semiarid zone of Gujarat is 23,319 km². The semiarid zone of Gujarat consists of deep black, medium black to loamy sand (goradu) type of soil. The mean annual rainfall over a semiarid zone varies from 600 mm to 1100 mm. The semiarid zone is geographically located between latitudes 72.314 to 74.476 and 23.457 to 21.817 longitudes. The index map of the semiarid zone of Gujarat is shown in Fig. 3.1.

3.2 Digital Elevation Model (DEM)

Figure 3.2 shows the digital elevation model for the semiarid region of Gujarat. The elevation ranges from −19 m to 820 m. The lower elevation range is near the coastal region of Anand district, while the higher elevation range is near the mountainous region of Pavaghar in Panchmahals district. The DEM is derived through the data of Shuttle Radar Topography Mission (SRTM) with a spatial resolution of 30 m × 30 m.

© The Author(s), under exclusive license to Springer Nature
Switzerland AG 2023
K. Naveena et al., *Spatio-temporal Trend Analysis of Rainfall using R Software and ArcGIS*, SpringerBriefs in Climate Studies,
https://doi.org/10.1007/978-3-031-48259-5_3

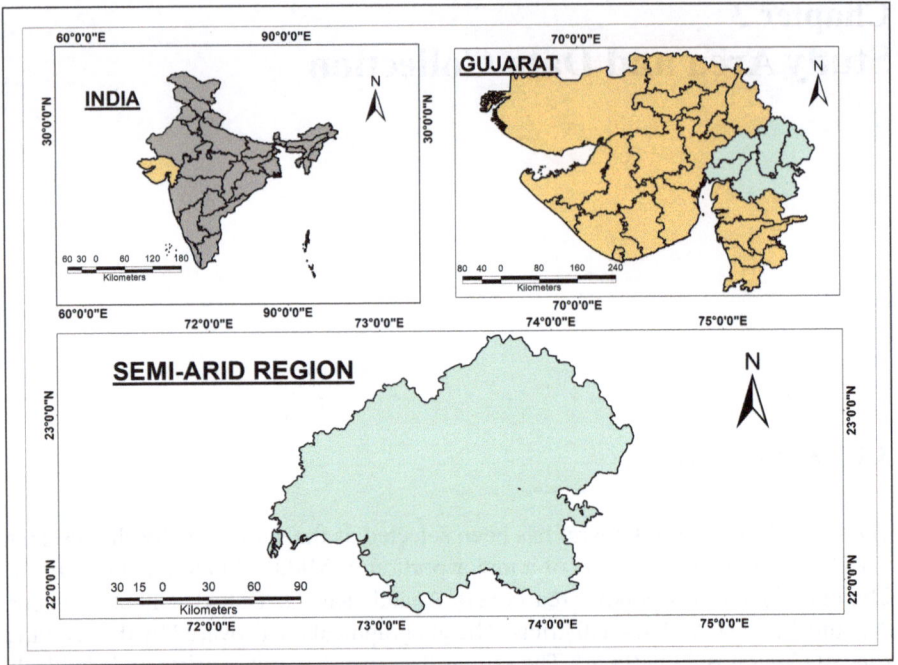

Fig. 3.1 Index map of the semiarid zone

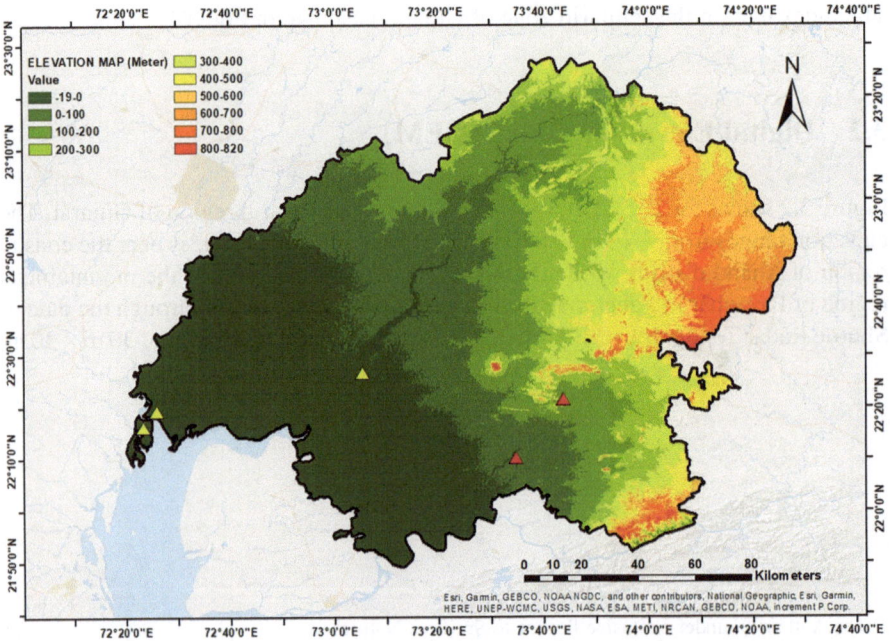

Fig. 3.2 Digital elevation model (DEM) of the semiarid zone

Table 3.1 Area and percentage area under different land use classes in the semiarid zone

Land use type	Area in km^2	Percentage of total area
Water	823.49	3.53
Trees	1209.001	5.1
Grass	7.33	0.03
Flooded vegetation	22.818	0.098
Agriculture	16,241	69.64
Scrub/shrub	2347	10.07
Built-up area	2630	11.28
Bare ground	39.46	0.16

3.3   Land Use/Land Cover Map (LU/LC)

The land use/land cover map has been prepared using the data of ESRI[1] in ArcGIS 10.3. The percentage area under each class is evaluated and represented in Table 3.1. The agricultural land has the highest percentage of area, 69.64%. The second largest percentage of area is the built-up area with 11.28% of the total area. The tree, grass, and scrubland comprises 15.2% of the total area. Further, it can be observed that a major portion of the built-up area is in the Vadodara and Kheda districts. Moreover, a major portion of the vegetation is observed in the eastern region of Dahod district and southern region of Chhota Udepur district (Fig. 3.3).

3.4   General Overview of the Study Area

The semiarid region covers a major portion of central Gujarat and part of the Charotar region. The mean rainfall over the study area is from 625 to 1000 mm. The type of soil in the study area is deep black, medium black to loamy sand (goradu). The major growing crops in the study area are cotton, pearl, tobacco, pulses, wheat, paddy, maize, and sugarcane.

3.5   Data Collection

The rainfall data has been collected through the State Water Data Centre (SWDC) in Gujarat and the Indian Meteorological Department (IMD) in Pune (Table 3.2).

---

[1] ESRI is an American multinational geographic information system (GIS) software company.

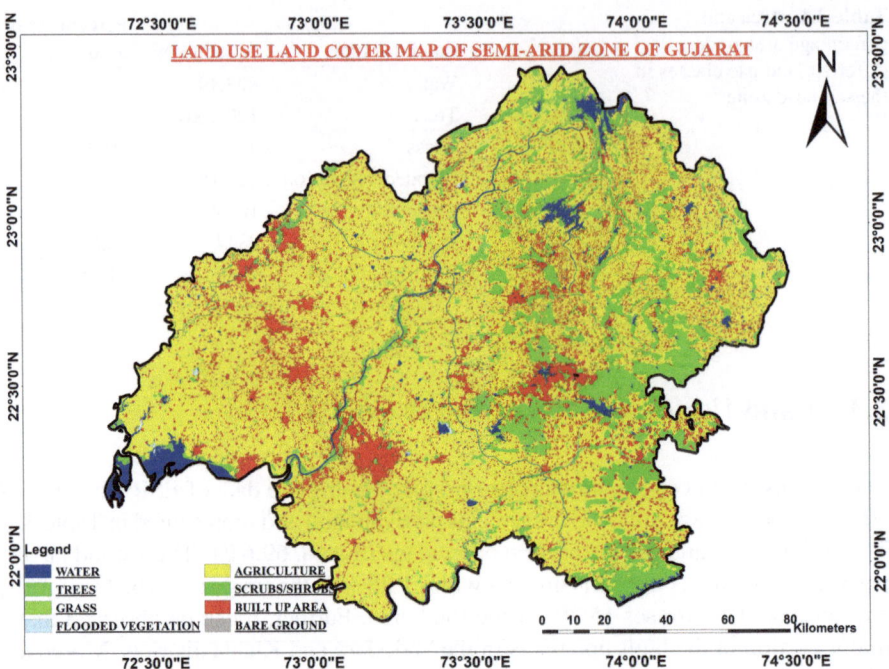

Fig. 3.3 Land use/land cover map of the semiarid zone

Table 3.2 Data source

Data	Description	Source
Topography (DEM)	30 m × 30 m resolution DEM	Shuttle Radar Topography Mission (STRM) 30 m resolution DEM through United States Geological Survey (USGS)
Rainfall data	Daily and monthly rainfall data	State water data Centre (SWDC), Gujarat Indian meteorological department (IMD), Pune
Land use/land cover maps	LU/LC maps of India for 1985, 1995, and 2005	DAAC (NASA)
Land use/land cover maps	LU/LC map for 2020	ESRI is an American multinational geographic information system (GIS) software company.

3.5.1   Daily Rainfall Data

The long-term daily rainfall data for 21 rain gauge stations has been provided by the State Water Data Centre (SWDC), Gujarat, and for 9 rain gauge stations, data has been provided by the Indian Meteorological department (IMD). The data has been taken over a span of 40–50 years. The latitude, longitude, mean rainfall, time span, and elevation of each station are shown in Table 3.3. The location of each rain gauge station is shown in Fig. 3.4.

Table 3.3 Details of the rain gauge stations

Sr. No	Station	District	Latitude	Longitude	Data From	Data To	No. of years	Elevation (m)
1	Borsad	Anand	22.2423	72.5321	1970	2012	42	0
2	Kanewal	Anand	22.2713	72.39	1970	2014	44	5.3
3	Siojitra	Anand	22.322	72.433	1970	2020	50	6.9
4	Kathlal	Kheda	22.5337	72.5953	1970	2020	50	15
5	Mehemdabad	Kheda	22.5007	72.4552	1970	2020	50	11
6	Dakor	Kheda	22.4456	73.0923	1970	2020	50	42
7	Limkheda	Dahod	22.83	73.99	1970	2020	50	200
8	Zalod	Dahod	23.09	74.16	1970	2020	50	269
9	C. Udepur	C. Udepur	22.3	74.01	1970	2020	50	150
10	Sankheda	C. Udepur	22.17	73.58	1970	2020	50	63
11	Naswadi	C. Udepur	22.04	73.73	1970	2010	40	81
12	Balasinor	Panchmahals	22.95	73.33	1970	2020	50	99
13	Shivrajpur	Panchmahals	22.2544	73.3546	1972	2011	41	39
14	Godhra	Panchmahals	22.77	73.61	1970	2013	43	124
15	Halol	Panchmahals	22.5	73.47	1970	2020	50	105
16	Kalol	Panchmahals	22.61	73.46	1970	2020	50	88
17	Jambughoda	Panchmahals	22.36	73.73	1970	2013	43	103
18	Shehra	Panchmahals	22.5658	73.3729	1970	2016	46	78
19	Karad dam	Panchmahals	22.23	73.44	1970	2013	43	45
20	Panam dam	Panchmahals	23.0327	73.4225	1974	2014	40	83
21	Morva	Panchmahals	22.54	73.57	1973	2013	40	103
22	Padra	Vadodara	22.24	73.08	1979	2020	41	30
23	Bhaniyara	Vadodara	22.2321	73.161	1977	2020	43	28
24	Karjan	Vadodara	22.0318	73.0642	1970	2019	49	24
25	Sinor	Vadodara	21.91	73.33	1970	2016	46	33
26	Savli	Vadodara	22.3256	73.1314	1970	2016	46	32
27	Waghodiya	Vadodara	22.1816	73.2413	1970	2010	40	31
28	Dabhoi	Vadodara	22.13	73.41	1970	2020	50	34
29	Vasad	Vadodara	22.45	73.06	1970	2010	40	44
30	Chandod	Vadodara	22	73.26	1970	2010	40	33

3.5.2   Specifications of LU/LC Maps

The LU/LC maps of the semiarid zone of Gujarat are prepared using the following data source:

1. *ESRI 2020 LU/LC*: The map is sourced from the European Space Agency (ESA) Sentinel-2 imagery at 10 m resolution. It comprises ten different classes of LU/LC. The details of the dataset are shown below.

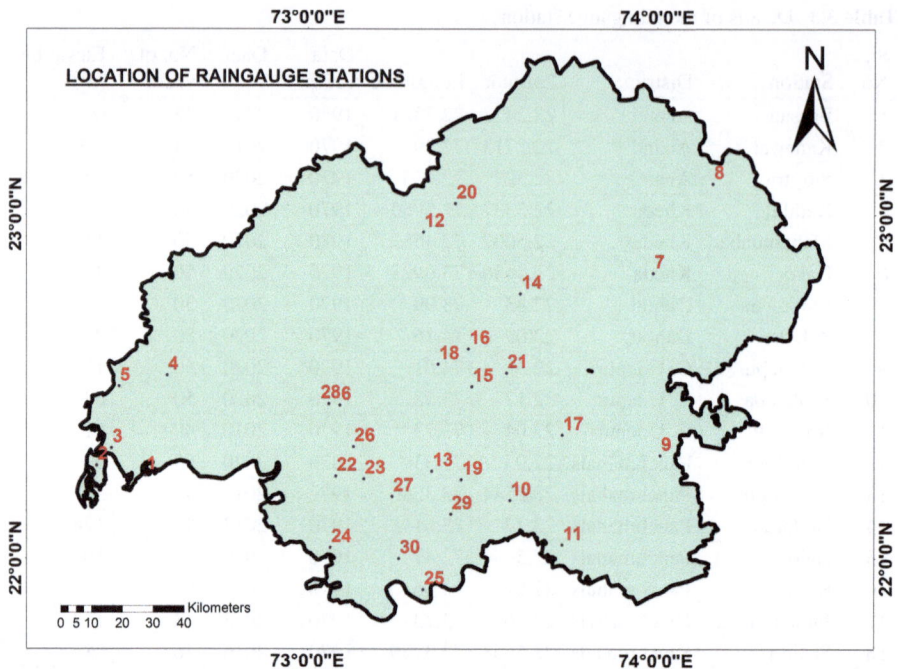

Fig. 3.4 Location of the rain gauge stations

Variable mapped: 2020 land use/land cover	
Mosaic projection: WGS84	
Source imagery: Sentinel-2	
Type: Thematic	
Source: Esri Inc.	
Publication date: July 2021	

2. *ORNL DAAC LU/LC map*: The LU/LC data at decadal intervals of 1985, 1995, and 2005 has been obtained from Oak Ridge National Laboratory Distributed Active Archive Centre (ORNL DAAC), which is NASA's earth observing system data and information system. The details of the dataset are shown below.

3.6   Data Classification

Spatial coverage: India (national boundaries conform to that published by the Survey of India).

Spatial resolution: 100 × 100 m.

Temporal resolution: Decadal.

Temporal coverage: 1 January 1985 to 31 December 2005 (Table 3.4).

Table 3.4 LU/LC classification as per International Geosphere-Biosphere Programme (IGBP) classification

Pixel value	Land use type (IGBP classification)	Description
1	Deciduous broadleaf forest	Woody vegetation having >60% area with height more than 2 m
2	Crop land	Temporarily cropped area
3	Built-up land	Covered by buildings and man-made structures
4	Mixed forests	Trees having a cover >60% and height >2 m.
5	Shrubland	Land with woody vegetation <2 m in height and with >10% shrub canopy cover
6	Barren land	Exposed soil, sand, rocks, or snow; do not have greater than 10% vegetated cover during the year
7	Fallow land	Cultivable land temporarily kept unused
8	Wasteland	These are land identified as currently underutilized and could be reclaimed to productive uses with reasonable effort. Degraded forest (<10% tree cover) with signs of erosion is classified under wasteland
9	Water bodies	Surface water: Fresh or saline in the form of ponds, lakes, reservoirs, rivers, etc.
10	Plantations	Commercial horticulture plantations; orchards and tree cash crops
11	Aquaculture	Land used to farm aquatic organisms including fish, mollusks, crustaceans, and aquatic plants
12	Mangrove forest	Evergreen forests in the intertidal areas. These forests are dense and dominated by halophytic plants
13	Salt pan	Land covered with salt and minerals
14	Grassland	Herbaceous types of cover. Tree and shrub cover is less than 10%
15	Evergreen broadleaf forest	Broadleaf woody vegetation with a cover >60% and height > 2 m. almost all trees and shrubs remain green year-round. Canopy is never without green foliage
16	Deciduous needleleaf forest	Woody vegetation with a percent cover >60% and height > 2 m. consists of seasonal needle leaf tree communities with an annual cycle of leaf-on and leaf-off periods
17	Permanent wetland	Land with permanent mixture of water and herbaceous or woody vegetation
18	Snow and ice	Land covered with snow or ice for most of the year
19	Evergreen needle forest	Needle leaf woody vegetation with a percent cover >60% and height exceeding 2 m. almost all trees remain green all year. Canopy is never without green foliage

Source: ORNL DAAC, https://daac.ornl.gov/

Chapter 4
Methodology

The detailed methodology used in the study is shown in Fig. 4.1.

4.1 Data Preparation

The arrangement of the SW monsoon and monthly (June, July, August, and September) rainfall is determined from the daily rainfall data collected at 30 rain gauge stations.

The missing rainfall data computation is carried out using the K-nearest neighbor method in R software.

4.2 Trend Detection Using Different Methods

The rainfall trend analysis has been carried out using various packages in R software.

4.2.1 Elbow Method to Determine Optimum Number of Clusters

The conventional approaches cannot be used to partition the data. Hence, it is necessary to use the approach that can automate the determination of the optimum numbers of clusters (Emrah and Dervis 2016). Among the different methods available, the elbow method is one of the best methods to determine the optimum number of clusters (Shi et al. 2021).

© The Author(s), under exclusive license to Springer Nature
Switzerland AG 2023
K. Naveena et al., *Spatio-temporal Trend Analysis of Rainfall using R Software and ArcGIS*, SpringerBriefs in Climate Studies,
https://doi.org/10.1007/978-3-031-48259-5_4

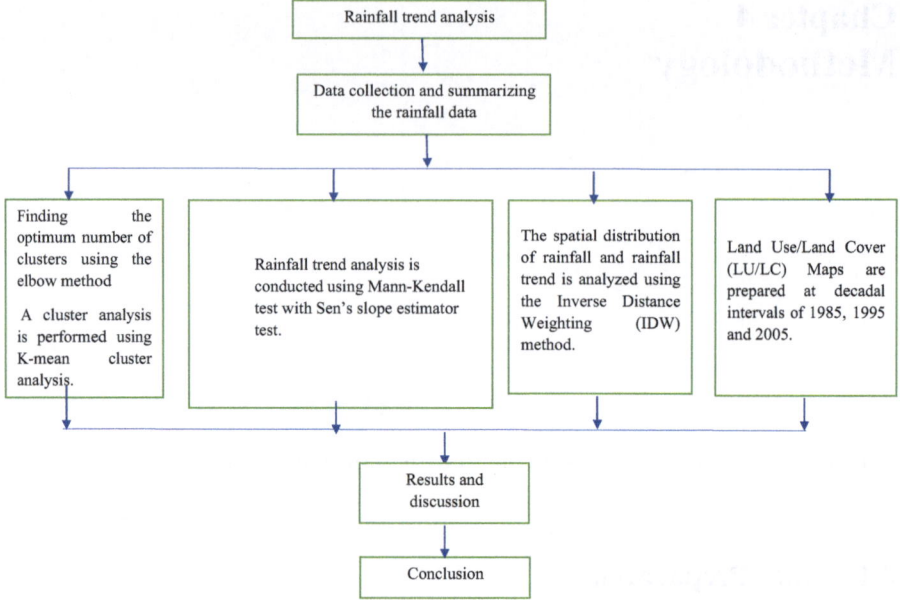

Fig. 4.1 A flowchart of the methodology

In this method, the k value is added one after another, and the sum of square errors (SSE) is noted where the SSE is the sum of average Euclidean distance of each point against the centroid.

When there is a sharp drop in the value of SSE and small angle formation, the value of k is obtained (Dhendra et al. 2018).

$$SSE = \sum_{k-1}^{k} \sum_{x=c_i} \left\| x - c_i \right\|^2 \tag{4.1}$$

where
x = The mean value for complete data
c_i = The mean value at a given station

4.2.2   Cluster Analysis

Conventional techniques usually utilized for analyzing rainfall patterns among different places are insufficient to model the variation of rainfall. So, we use cluster analysis, and the first step is to obtain the optimum number of clusters in the observed rainfall data. Then, the rainfall data are divided into the required number of clusters (Kannan and Ghosh 2011).

K-means clustering aims to classify n observations $x = \{x_1, x_2, x_3, \cdots x_n\}$ into k clusters $C = \{C_1, C_2, \cdots C_k\}$, and the objective function is defined as the total distance between all observations from their respective cluster centers. The main objective of k-means is to minimize this within-cluster sum of squares (WCSS).

$$\text{WCSS} = C^{\text{ar gmin}} \sum_{i=1}^{k} \sum_{x \, c_i} \left\| x - c_i \right\|^2 \tag{4.2}$$

where
x = The mean value for complete data
c_i = The mean value at a given station

The two main steps of cluster analysis are the distribution of data among clusters and updating the center of clusters. The algorithm is alternated between the two steps until the value of an objective function cannot be further reduced. K-means is a hard clustering method in which each observation belongs to one cluster only (Li 2017).

4.2.3   Mann-Kendall Test

To find out statistically significant trends in various climatical parameters such as rainfall, stream flow, and temperature and nonparametric tests such as the Mann-Kendall (MK) test (Mann 1945; Kendall 1975; Gilbert 1987) have been broadly used by researchers (Jain et al. 2012). Mann-Kendall test is used to identify the monotonic upward or downward trend (Zinabu and Dioha 2020).

The basic requirement to use the Mann-Kendall test is that the data should not be auto-correlated (Majed et al. 2021). In the analysis, the null hypothesis states that there is no trend in the data in which the population is identically distributed. The alternate hypothesis states that data has a monotonic trend. This method has the advantage of the data not being required to be normally distributed as the method is a nonparametric test (Ali and Bahruddin 2021).

$$S = \sum_{j=1}^{m-1} \sum_{k=j+1}^{m} \text{sgn}\left(f_j - f_k \right) \tag{4.3}$$

where m = length of time series

The test can be applied for f_j rank from $j = 1, 2, 3, \ldots, m - 1$ to f_k from $k = j + 1$, $j + 2, \ldots, m$.

$sgn(f_j - f_k)$ can be given as:

$$sgn\left(f_j - f_k\right) = 1, if\left(f_j - f_k\right) > 0$$

$$= 0, if\left(f_j - f_k\right) = 0$$

$$= -1, if\left(f_j - f_k\right) < 0 \qquad (4.4)$$

$$Var_s = \frac{m\left(m-1\right)\left(2m+5\right) - \sum_{q=1}^{k} t_k\left(t_k - 1\right)\left(2t_k + 5\right)}{18} \qquad (4.5)$$

In case when $m > 10$, standard normal test statistics is calculated as:

$$Z = \begin{cases} \dfrac{S-1}{\sqrt{Var_s}}, if\ S > 0 \\ 0, \quad if\ S > 0 \\ \dfrac{S+1}{\sqrt{Var_s}}, if\ S < 0 \end{cases} \qquad (4.6)$$

Here, the Z value denotes the direction of trend whether positive or negative. The positive sign of Z indicates a positive trend, while the negative sign indicates a negative trend.

4.2.4 Sen's Slope Estimator Test

The magnitude of the slope can be calculated using a nonparametric test (Gajbhiye et al. 2016) called Sen's slope estimator test, which was given by Sen in 1968. The Sen's slope value (β) in the equation represents the trend, and the value indicates the steepness of the trend (Gocic and Trajkovic 2012). This approach to estimating the slope of the trend is better than the linear regression method as it limits the effect of outlining the values of data (Suryavanshi et al. 2014).

$$Sen's\ slope\left(\beta\right) = Median\frac{\left(f_j - f_k\right)}{\left(j - k\right)} \qquad (4.7)$$

where
f_j and f_k are the value at time of j and k, respectively

4.2.5 Spatial Distribution Using the Inverse Distance Weighting (IDW) Method

The inverse distance weighting method is based on the first law of geography given by Tobler (1970). It may be stated that each thing is relatable to another thing, but the nearby things are more relatable as compared to distant things. The US National Weather Service developed the IDW method in 1972 (Feng-Wen and Wuing Liu 2012). As per the IDW method, a parameter at a neighboring point influences more compared to one at a farther point, and the effect of the known data point is inversely proportional to the distance between known and unknown data points (Bhargava et al. 2016). The IDW method is a deterministic method and is very useful in analyzing the variability, patterns, and concentration of geo-reference variables (Biswas et al. 2020).

In this method, the weightage factor is given by the equation:

$$W_i = \frac{D_k^{-y}}{\sum_{k=1}^{m} D_k^{-y}} \tag{4.8}$$

where
W_i = Weightage of each rain gauge station
D_k = Distance of a given rain gauge station to an unknown point

4.3 Introduction to "R" Software

Advantages/Capabilities of R Software An effective data handling and storage facility.

- A suite of operators for calculation of large arrays.
- Graphic capabilities for data analysis.
- Results can be extracted in tabular format, and the graphs can also be exported in PDF or JPEG format.

The different windows in R software are shown in Fig. 4.2.
The functions of these windows are as follows:

1. *Command window*: In the command window, the programming part is to be carried out.
2. *Result window*: When we run the analysis, and the processing is done, the results in terms of various statistical parameters will be shown in this window.
3. *History window*: This part of the screen shows the history of the commands performed.

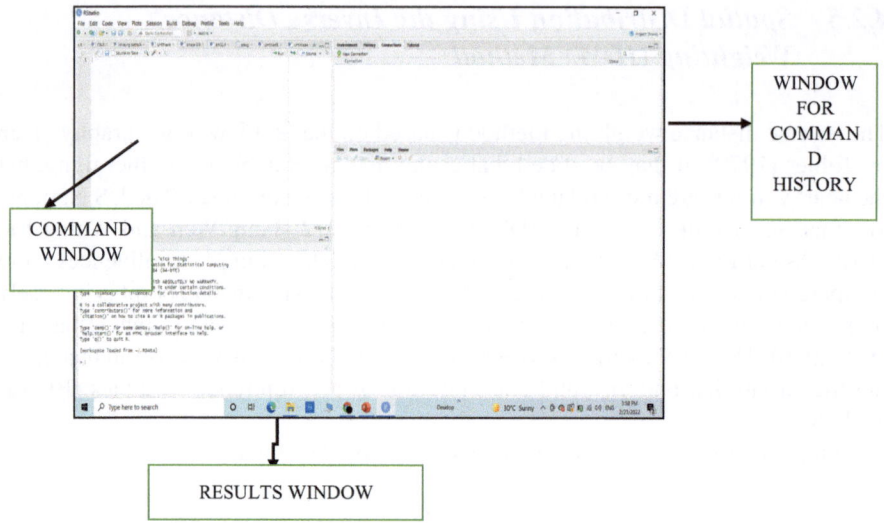

Fig. 4.2 Different windows in R software

Table 4.1 Various packages used in R software

Sr.no	Method	Package	Command
1	K-NN method	Visualization and Imputation of Missing Values (VIM)	Knn
2	Elbow method	Nbclust	Nbclust
3	K-means	Factoextra	Km.Res
4	Mann-Kendall test	Trend	Mk.Test
5	Sen's slope estimator test	Trend	Sens.Slope

4.4 Various Packages Used in R Software

The various packages used in R software are shown in Table 4.1.

4.5 Generalized Procedure to be Followed in R Software

1. Give the extension to the given script in the desired folder.
2. Prepare the monthly rainfall data table of a given station in Excel format.
3. Input the data in R software using command read.table.
4. Install the "Trend" package from the toolbar.
5. Bring the installed package to the library of R software.
6. Perform the trend analysis using MK test and Sen's .slope estimator test command.
7. Note down the results.

4.6   R Code for Rainfall Trend Analysis

```
# Load the required packages
library(trend)
library(modifiedmk)
library(trendchange)
# Load rainfall data from clipboard (ensure data is properly formatted)
rainfall <- read.table(file = "clipboard", header = TRUE)
# Mann-Kendall Test for Trend
mk_test_result <- mk.test(x = rainfall, alternative = "two.sided")
# Calculate the magnitude of trend using Sen's slope
sens_slope_result <- sens.slope(x = rainfall, conf.level = 0.95)
# Seasonal Mann-Kendall Test for Trend
rainfall <- ts(rainfall, start = c(1921, 1), frequency = 12)
smk_test_result <- smk.test(rainfall)
# Apply Wallis and Moore Phase-Frequency Test for Randomness
wm_test_result <- wm.test(rainfall)
```

Different Modified Mann-Kendall Test under "modifiedmk" package

```
# Modified Mann-Kendall Test for Serial Correlation using Hamed and Ramachandra
Rao (1998) Approach
mmkh_result <- mmkh(x = rainfall)
# Apply Mann-Kendall Trend Test Applied to Trend-Free Prewhitened Time
Series Data
# in Presence of Serial Correlation using Yue and Wang (2002) Approach
tfpwmk_result <- tfpwmk(x = rainfall)
# Apply Modified Mann-Kendall Test for Serially Correlated Data using Yue and
Wang (2002)Approach
mmky_result <- mmky(x = rainfall)
# Bootstrapped Mann-Kendall Trend Test with Optional Bias Corrected Prewhitening
(Hamed 2009)
pbmk_result <- Pbmk(x = rainfall)
# Innovative Trend Analysis Method proposed by Sen (2012)
innovtrend_result <- innovtrend(x = rainfall, ci = 95)
```

References

Bhargava N, Bhargava R, Tanwar PS, Narooka (2016) Comparative study of inverse power of IDW interpolation method in inherent error analysis of aspect variable, vol 2. https://doi.org/10.1007/978-981-10-3920-1

Biswas RN, Islam N, Mia J (2020) Modeling on the spatial vulnerability of lightning disaster in Bangladesh using GIS and IDW techniques. Spat Inf Res 28:507–521. https://doi.org/10.1007/s41324-019-00311-y

Dhendra M, Sunarna H, Ekaprana W, Muljono (2018) The determination of cluster number at k-mean using elbow method and purity evaluation on headline news. International Seminar on Application for Technology of Information and Communication

Emrah H, Dervis K (2016) A comprehensive survey of traditional, merge-split and evolutionary approaches proposed for determination of cluster number. Swarm Evolution Computat. https://doi.org/10.1016/j.swevo.2016.06.004

Gajbhiye S, Meshram C, Mirabbasi R (2016) Trend analysis of rainfall time series for Sindh river basin in India. Theor Appl Climatol 125:593–608. https://doi.org/10.1007/s00704-015-1529-4

Gocic M, Trajkovic S (2012) Analysis of changes in meteorological variables using Mann-Kendall and Sen's slope estimator statistical tests in Serbia. Glob Planet Chang 100:172–182. https://doi.org/10.1016/j.gloplacha.2012.10.014

Hamed KH (2009) Enhancing the effectiveness of prewhitening in trend analysis of hydrologic data. J Hydrol 368:143–155. https://doi.org/10.1016/j.jhydrol.2009.01.040

Hamed KH, Ramachandra Rao A (1998) A modified Mann-Kendall trend test for autocorrelated data. J Hydrol 204(1–4):182–196. https://doi.org/10.1016/S0022-1694(97)00125-X

Jain SK, Kumar V, Saharia M (2012) Analysis of rainfall and temperature trend in Northeast India. Int J Climatol. https://doi.org/10.1002/joc.3483

Kannan S, Ghosh S (2011) Prediction of daily rainfall state in a river basin using statistical downscaling from GCM output. Stoch Environ Res Risk Assessment 25:457–474. https://doi.org/10.1007/s00477-010-0415-y

Kendall (1975) Rank correlation methods charles griffin, London, pp 34–37

Li J (2017) Clustering and forecasting for rain attenuation time series data, Stockholm, Sweden, p 11. https://doi.org/10.1016/j.jeconom.2020.06.008

Majed A, Kumari M, Mallic M, Ramakrishnan R, Islam S, Kumar C (2021) Time series trend analysis of rainfall in last five decades and its quantification in Aseer region of Saudi Arabia. Arab J Geosci 14:519. https://doi.org/10.1007/s12517-021-06935-5

Mann HB (1945) Nonparametric tests against trend. Econometrica 13:245–259. https://doi.org/10.2307/1907187

Sen Z (2012) Innovative trend analysis. J Hydrol Eng 17(9):1042–1046. https://doi.org/10.1061/(ASCE)HE.1943-5584.0000556

Shi C, Wei B, Wei S et al (2021) A quantitative discriminant method of elbow point for the optimal number of clusters in clustering algorithm. Wireless Com Network 31. https://doi.org/10.1186/s13638-021-01910-w

Suryavanshi S, Pandey A, Chaube UC et al (2014) Long-term historic changes in climatic variables of Betwa Basin, India. Theor Appl Climatol 117:403–418. https://doi.org/10.1007/s00704-013-1013-y

Yue S, Wang CY (2002) Applicability of prewhitening to eliminate the influence of serial correlation on the Mann-Kendall test. Water Resour Res 38(6):41–47. https://doi.org/10.1029/2001WR000861

Zinabu A, Dioha MO (2020) Climate change and trend analysis of temperature: the case of Addis Ababa, Ethiopia. Environ Syst Res 9(27). https://doi.org/10.21203/rs.3.rs-41363/v1

Chapter 5
Computations

5.1 Cluster Analysis

To detect the significant differences between the mean rainfall of two clusters, an independent sample t-test at a significance level of $p = 0.05$ is carried out in Excel. The result of the independent t-test shows that the mean rainfall of two clusters differs significantly at $p = 0.021$. A difference of more than 200 mm rainfall is observed between the two groups, with the first group showing more rainfall than the second group (Fig. 5.1).

Then, the semiarid zone is divided into two clusters having 13 and 17 stations. The mean rainfall distribution of cluster 1 is 935.10 ± 348.17 mm/year and cluster 2 is 76.512 ± 266.00 mm/year.

Results show that 60% of the stations from Panchmahals district and 100% of the stations of Chhota Udepur district come under the first cluster, while 66% of Vadodara as well as all stations of Anand and Kheda districts come under the second cluster. The stations under the first cluster show a higher magnitude of mean rainfall than the second cluster.

5.2 Rainfall Trend Analysis of SW Monsoon

5.2.1 Mann-Kendall and Sen's Slope Estimator Test

Rainfall trend analysis of 30 rain gauge stations has been carried out using the parametric Mann-Kendall test for 30 rain gauge stations. The results are shown in Table 5.1.

© The Author(s), under exclusive license to Springer Nature
Switzerland AG 2023
K. Naveena et al., *Spatio-temporal Trend Analysis of Rainfall using R Software and ArcGIS*, SpringerBriefs in Climate Studies,
https://doi.org/10.1007/978-3-031-48259-5_5

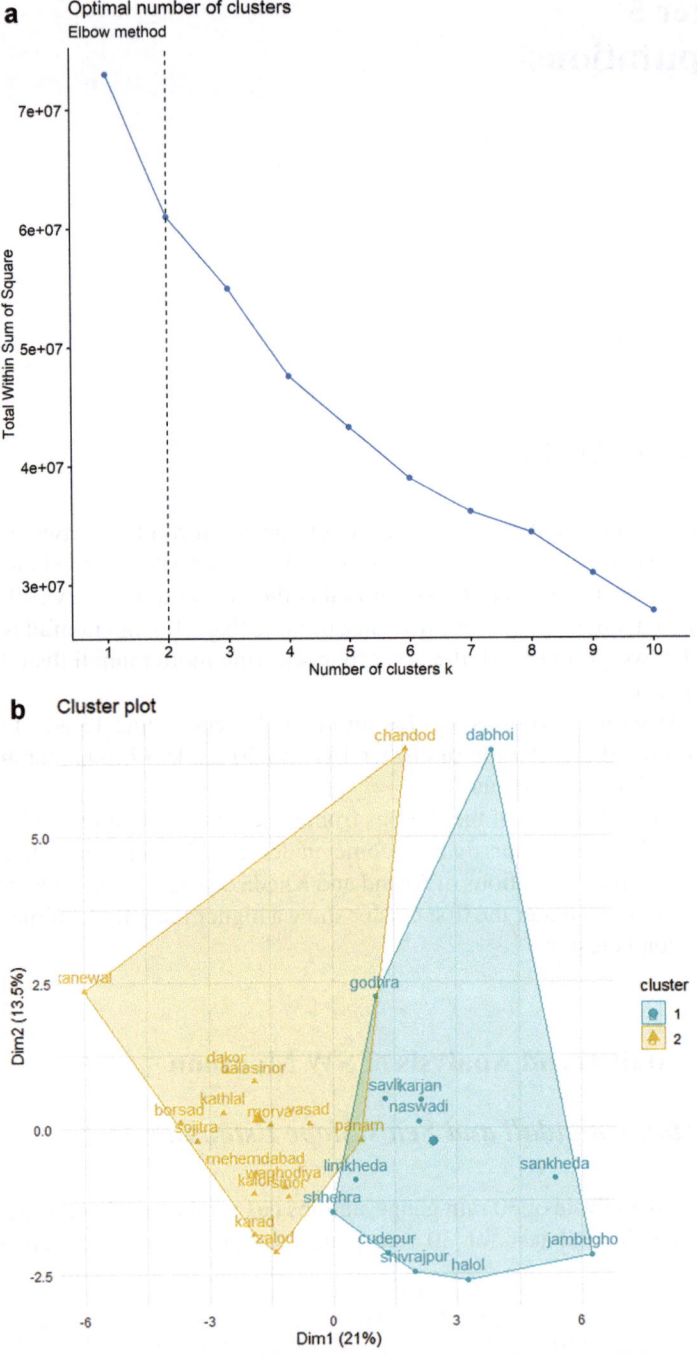

Fig. 5.1 Cluster analysis plot of the study area. (**a**) Elbow method. (**b**) Cluster analysis using the K-mean method. (**c**) Cluster map

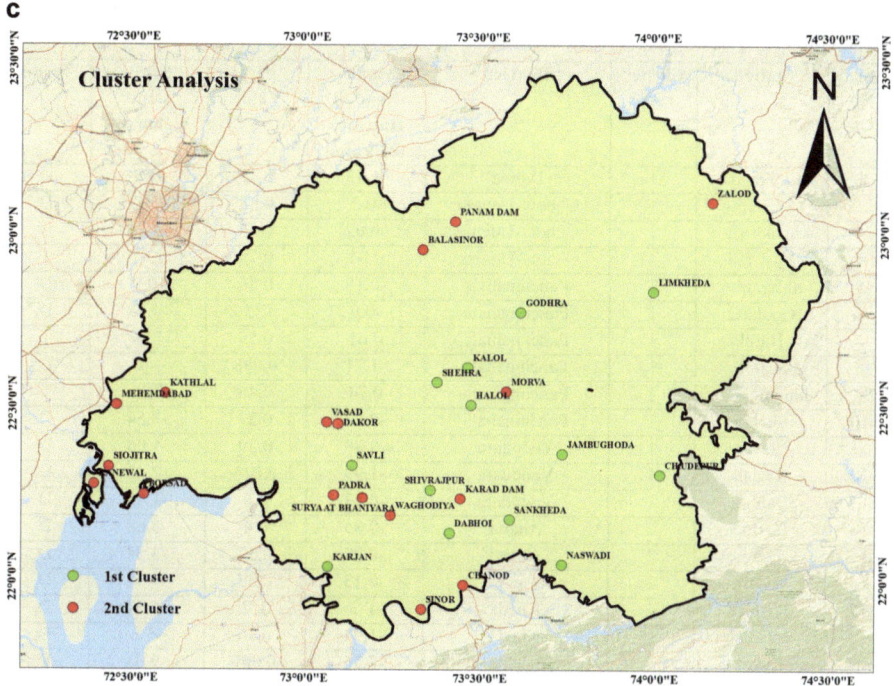

Fig. 5.1 (continued)

Table 5.2 summarizes the results of the SW monsoon rainfall trend analysis using the Mann-Kendall test with Sen's slope estimator test. Here, the positive sign indicates an increasing trend, whereas the negative sign indicates a decreasing trend. The results show three stations with a significant decreasing trend. From the results, it has been observed that 20 stations have shown a decreasing trend and 10 stations have shown an increasing trend.

5.2.2   Spatial Distribution of SW Monsoon Rainfall and Rainfall Trend Using IDW Method

The spatial variability of SW monsoon rainfall and rainfall trend has been analyzed using the inverse distance weighting (IDW) method. For analyzing rainfall variability, the mean SW monsoon rainfall is taken as the data for each station, whereas the mean Sen's slope value based on the Mann-Kendall test is used to detect the spatial variability of rainfall trend.

The analysis is shown in Fig. 5.2.

The spatial distribution map of SW monsoon rainfall shows lower mean rainfall near the coastal region (Anand and Kheda districts), while a higher range of rainfall

Table 5.1 Results of Mann-Kendall test with Sen's slope estimator test for SW monsoon rainfall data

Sr. No	Station	Cluster rank	District	Mann-Kendall statistics 'Z' Value	'P' Value	Sen's slope Value (β)	Trend
1	Limkheda	1	Dahod	-1.43	0.15	-4.9	D
2	C. Udepur	1	Chhota-Udepur	-0.2	0.85	-0.93	D
3	Sankheda	1	Chhota-Udepur	-0.62	0.53	-2.76	D
4	Naswadi	1	Chhota-Udepur	0.32	0.75	1.67	I
5	Shivrajpur	1	Panchmahals	-0.18	0.86	-0.9	D
6	Godhara	1	Panchmahals	-0.9	0.356	-4.38	D
7	Halol	1	Panchmahals	0.61	0.55	1.89	I
8	Kalol	1	Panchmahals	-1.71	0.088	-4	D
9	Jambughoda	1	Panchmahals	0.74	0.46	3.38	I
10	Shehra	1	Panchmahals	-0.88	0.37	-3.54	D
11	Savli	1	Vadodara	-0.88	0.39	-5.06	D
12	Karjan	1	Vadodara	-1.99	**0.046**	-7.43	D
13	Dabhoi	1	Vadodara	-0.53	0.6	-1.86	D
14	Borsad	2	Anand	-0.85	0.4	-3.68	D
15	Kanewaal	2	Anand	-0.89	0.37	-3.54	D
16	Sojitra	2	Anand	-0.13	0.89	-0.38	D
17	Kathlal	2	Kheda	-0.3	0.76	-1.51	D
18	Mehemdabad	2	Kheda	-1.94	0.13	-5.6	D
19	Dakor	2	Kheda	-2.76	0.007	-9.03	D
20	Zalod	2	Dahod	-2.06	0.039	-6.7	D
21	Balasinor	2	Panchmahals	1.1293	0.26	3.7	I
22	Panam Dam	2	Panchmahals	0.69	0.5	3.57	I
23	Morva	2	Panchmahals	0.3	0.76	1.25	I
24	Karad Dam	2	Panchmahals	-0.11	0.92	-1.18	D
25	Chandod	2	Vadodara	1.25	0.21	6.6	I
26	Sinor	2	Vadodara	-1.96	0.050	.-7.33	D
27	Waghodiya	2	Vadodara	-0.97	0.33	-5.77	D
28	Vasad	2	Vadodara	0.57	0.56	3.31	I
29	Padra	2	Vadodara	-0.99	0.32	-5	D
30	Bhaniyara	2	Vadodara	0.87	0.38	3.67	I

is observed near the central and southeast regions (Chhota Udepur and part of Panchmahals district).

The red and yellow triangle shows the stations with the highest and lowest mean SW monsoon rainfall in the semiarid region.

The highest mean SW monsoon is observed at Jambughoda station, while the lowest is observed at Kanewal station (Fig. 5.3).

The spatial distribution map of SW monsoon rainfall trend shows a higher increasing trend near the central and southern regions (Chhota Udepur district). The rainfall trend shows a higher decreasing trend near the southwest region (Vadodara district).

Table 5.2 Results of Mann-Kendall test with Sen's slope estimator test for monthly rainfall data

Sr. No	Station	Cluster Rank	Month	Mann-Kendall statistics 'Z' Value	'P' Value	Sen's slope value	Trend
1	Limkheda		June	-1.77	0.078	-1.3	D
			July	-0.58	0.56	-0.71	D
			August	-1.03	0.3	-2.17	D
		1	September	0.53	0.6	0.27	I
2	C. Udepur		June	-1.9	0.06	-1.65	D
			July	-0.1	0.9	-0.284	D
			August	-0.15	0.89	-0.28	D
		1	September	0.52	0.6	0.62	I
3	Sankheda		June	-1.67	0.1	-1.5	D
			July	-0.53	0.6	-0.89	D
			August	-0.63	0.52	-1.04	D
		1	September	0.43	0.67	0.63	I
4	Naswadi		June	-0.83	0.4	-1.14	D
			July	0.97	0.34	1.83	I
			August	-1.24	0.22	-2.92	D
		1	September	0.68	0.5	1.26	I
5	Shivrajpur	1	June	-0.33	0.744	-0.33	D
			July	-0.093	0.93	-0.26	D
			August	-0.67	0.51	-2.31	D
			September	0.26	0.79	0.28	I
6	Godhara		June	-1	0.28	-0.95	D
			July	-0.58	0.56	-0.56	D
			August	-0.967	0.33	-2.56	D
		1	September	0.28	0.78	0.27	I
7	Halol		June	-1.2	0.23	-1	D
			July	0.89	0.37	1.5	I
			August	0.36	0.73	0.85	I
		1	September	1	0.32	1	I
8	Kalol		June	-2.308	**0.021**	-1.5	D
			July	-0.35	0.73	-0.33	D
			August	-1.583	0.11	-2.13	D
		1	September	1.17	0.24	0.82	I
9	Jambughoda		June	-1.6	0.099	-2.31	D
			July	0.26	0.79	0.52	I
			August	0.612	0.54	2.15	I
		1	September	0.96	0.34	1.32	I
10	Shehra		June	-2.1	**0.03**	-0.33	D
			July	-0.34	0.74	-0.54	D
			August	-1.07	0.28	-1.92	D
		1	September	1.009	0.31	1.17	I
11	Savli		June	-2	**0.05**	-2	D
			July	-0.03	0.97	-0.03	D
			August	-1.5	0.15	-3.6	D
		1	September	0.8	0.43	0.65	I

Table 5.2 (continued)

Sr. No	Station	Cluster Rank	Month	Mann-Kendall statistics 'Z' Value	'P' Value	Sen's slope value	Trend
12	karjan		June	-0.87	0.38	-1	D
			July	-0.1	0.92	-0.21	D
		1	August	-2.53	**0.011**	-3.88	D
			September	-0.07	0.94	-0.052	D
13	Dabhoi		June	-2	**0.04**	-1.71	D
			July	-0.58	0.56	-0.7	D
			August	-0.48	0.63	-0.62	D
		1	September	1.39	0.16	1.55	I
14	Borsad		June	-1.83	0.066	-1.87	D
			July	0.21	0.84	0.5	I
			August	-0.48	0.63	-1.21	D
		2	September	-0.65	0.52	-0.35	D
15	Kanewaal		June	-0.24	0.8	0	D
			July	2.06	**0.039**	3.62	I
			August	1.58	0.11	2	I
		2	September	1.88	0.06	1.6	I
16	Sojitra		June	-1.58	0.11	-1.08	D
			July	0.5	0.6	0.55	I
			August	0.59	0.55	1.31	I
		2	September	0.93	0.35	0.33	I
17	Kathlal		June	-0.76	0.45	-0.5	D
			July	1.1	0.28	1.62	I
			August	-0.2	0.84	-0.39	D
		2	September	-0.31	0.76	0	D
18	Mehemdabad		June	-3.76	0.00016	-2.67	D
			July	-0.53	0.59	-0.83	D
			August	-0.54	0.58	-0.9	D
		2	September	1.24	0.21	1.03	I
19	Dakor		June	-2.44	0.015	-1.67	D
			July	-0.93	0.35	-1.63	D
		2	August	-1.95	0.051	-3.88	D
			September	-0.091	0.93	-0.008	D
20	Zalod		June	-2.8	0.005	-2.3	D
			July	-0.85	0.4	-0.96	D
			August	-1.0236	0.31	-1.76	D
		2	September	-0.67	0.5	-0.6	D
21	Balasinor		June	-0.97	0.33	-0.47	D
			July	1.1584	0.25	1.55	I
			August	0.25	0.81	0.58	I
		2	September	0.87	0.38	0.6	I
22	Panam Dam		June	-0.081	0.94	-0.068	D
			July	1.25	0.21	3	I
			August	-0.59	0.56	-1.9	D
		2	September	1.26	0.21	1.8	I

Table 5.2 (continued)

Sr. No	Station	Cluster Rank	Month	Mann-Kendall statistics 'Z' Value	'P' Value	Sen's slope value	Trend
23	Morva		June	-0.51	0.61	-0.36	D
			July	0.62	0.52	1.55	I
			August	0.078	0.93	0.17	I
		2	September	1.3	0.2	1.42	I
24	Sinor		June	-3.23	0.0012	-3.075	D
			July	0.24	0.8	0.4	I
			August	-2.56	0.012	-4.92	D
		2	September	0.92	0.36	1.03	I
25	Waghodiya		June	-0.46	0.65	-0.5	D
		2	July	-0.49	0.62	-1.35	D
			August	-0.88	0.38	-2.61	D
			September	-0.35	0.73	-0.3	D
26	Vasad		June	0.49	0.62	0.33	I
			July	1.33	0.18	4	I
			August	-0.22	0.83	-0.2	D
		2	September	-0.088	0.93	0	D
27	Karad Dam		June	-1.48	0.16	-1.27	D
			July	-0.25	0.81	-0.4	D
			August	-0.2	0.84	-0.69	D
		2	September	1.2356	0.22	1.88	I
28	Padra	2	June	-1.9	0.04	-1.67	D
			July	-0.99	0.32	-1.72	D
			August	-1.16	0.25	-1.9	D
			September	1.44	0.15	1.13	I
29	Bhaniyara	2	June	-2.42	0.015	-2.366	D
			July	0.98	0.32	2.13	I
			August	0.45	0.65	0.87	I
			September	1.55	0.12	1.59	I
30	Chandod		June	0.2	0.83	0.2	I
			July	2.11	**0.03**	4.73	I
			August	-0.3	0.76	-0.56	D
		2	September	0.83	0.39	0.64	D

5.3   Monthly Rainfall Trend Analysis

5.3.1   Mann-Kendall Test and Sen's Slope Estimator Test

The monthly trend analysis of 30 rain gauge stations using the Mann-Kendall test and Sen's slope estimator test has been indicated in Table 5.2.

Results of Mann-Kendall test (Table 5.2) indicate that in June, about ten stations show a significant negative trend; in July, only two stations show a significant negative trend; and in August, two stations show a significantly decreasing trend.

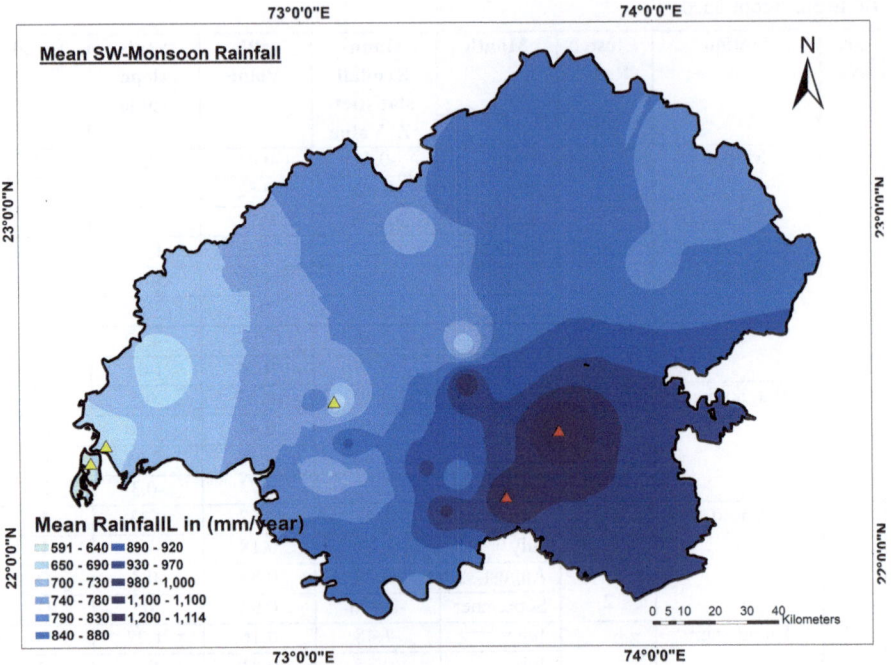

Fig. 5.2 Spatial distribution map of SW monsoon rainfall

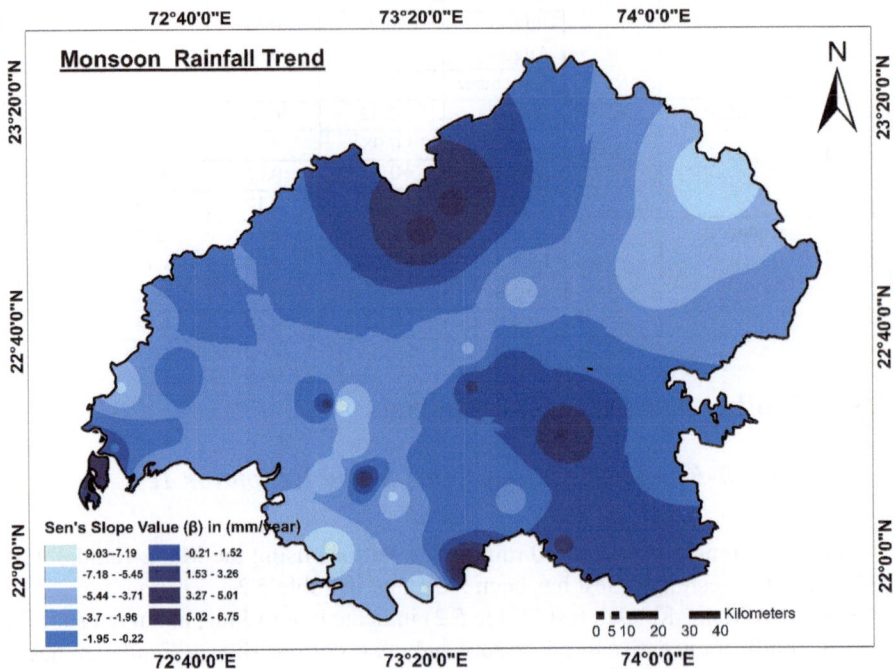

Fig. 5.3 Spatial distribution map of SW monsoon rainfall trend

The spatial distribution maps of the mean rainfall and rainfall trend (Sen's slope value) station are shown in Fig. 5.4 (a–d). From the spatial distribution maps of mean rainfall, it is observed that the higher mean rainfall occurs over the southern region of the Chhota Udepur district, while the lower mean rainfall is observed near the western region of the Anand and Kheda districts for all the months. The spatial distribution maps of Sen's slope estimator test show the highest values near the western part of the Anand district, while the lower values occur over the eastern region.

5.4   Spatial Distribution Maps of Monthly Mean Rainfall and Mean Rainfall Trend

Spatial distribution of the mean monthly rainfall is shown in Fig. 5.4. From the figure, it can be seen that higher rainfall is observed near central Gujarat and Chhota Udepur district, while lower rainfall is observed near the coastal region of Anand and Kheda districts.

The spatial distribution maps of the mean rainfall trend are shown in Fig. 5.5 (a–d). From the figure, it can be seen that in the months of June, August, and September, a higher rainfall trend is observed near central Gujarat and Chhota Udepur district, while in August, a higher rainfall trend is observed near the coastal region of Anand and Kheda districts.

5.5   Effect of Rainfall Trend on Land Use/Land Cover Pattern

From the trend analysis results, it is observed that four stations show a significant decreasing trend in the monsoon, of which two stations belong to the Vadodara district. In June, 10 out of the 30 stations show a significant decreasing trend, out which 5 are from the Vadodara district. In August, two stations show a significant decreasing trend and both stations are from the Vadodara district. In June, July, and August, the strongest negative trend is observed at Padra and Sinor stations, which belong to the Vadodara district.

From all these results, it can be observed that out of all the significant decreasing results of the semiarid region, most of the results come from the Vadodara district. To link and relate the changes in the rainfall pattern with land use/land cover (LU/LC) patterns, the LU/LC maps are prepared at decadal intervals of 1985, 1995, and 2005 as shown in Fig. 5.6. The results of the analysis are shown in Table 5.3. The maps show that the built-up area significantly increased in the Vadodara district between 1985 and 2005, which may be a reason for the decreasing rainfall trend in the Vadodara district.

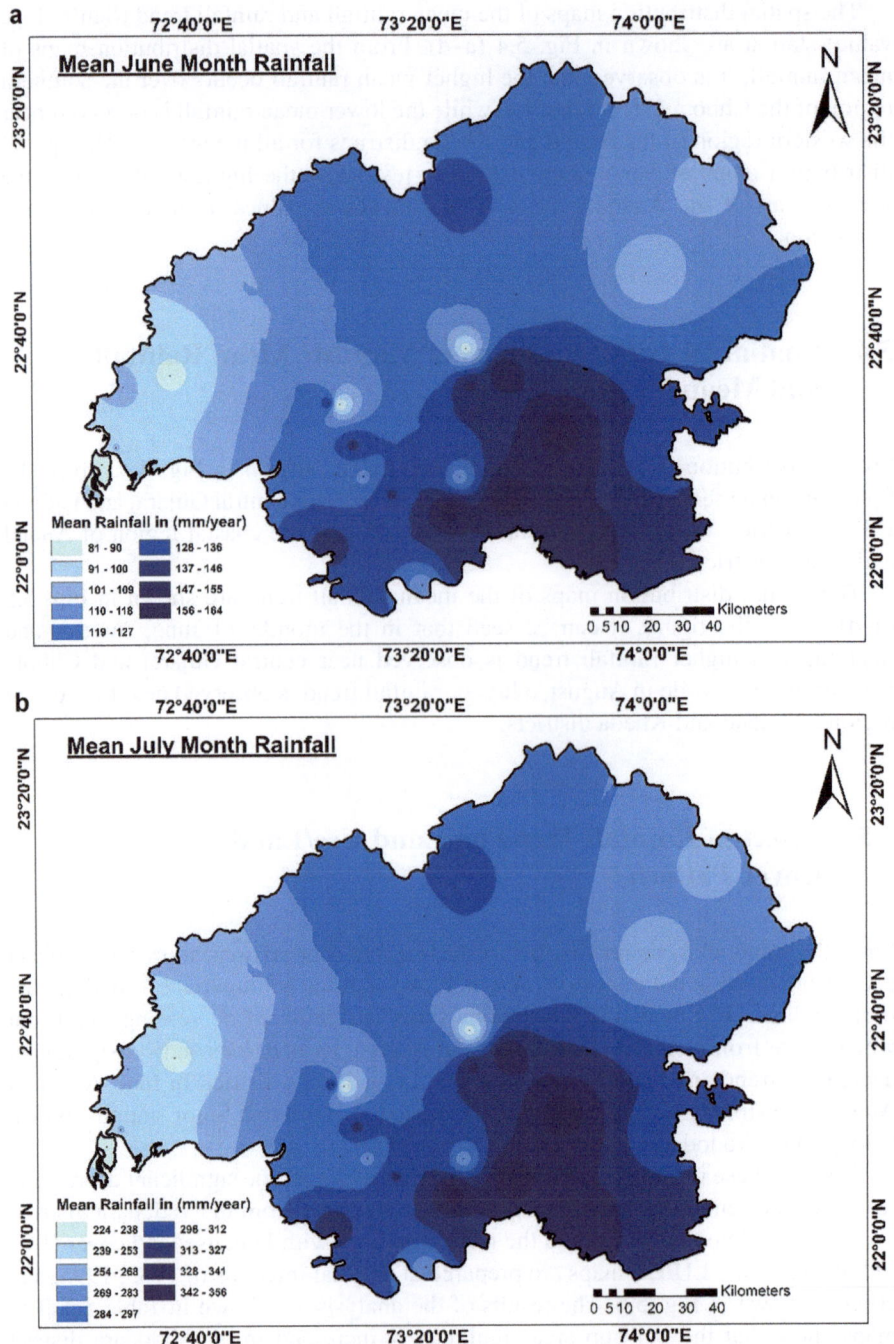

Fig. 5.4 (**a**) Spatial distribution maps of mean monthly rainfall in June. (**b**) Spatial distribution maps of mean monthly rainfall in July. (**c**) Spatial distribution maps of mean monthly rainfall in August. (**d**) Spatial distribution maps of mean monthly rainfall in September

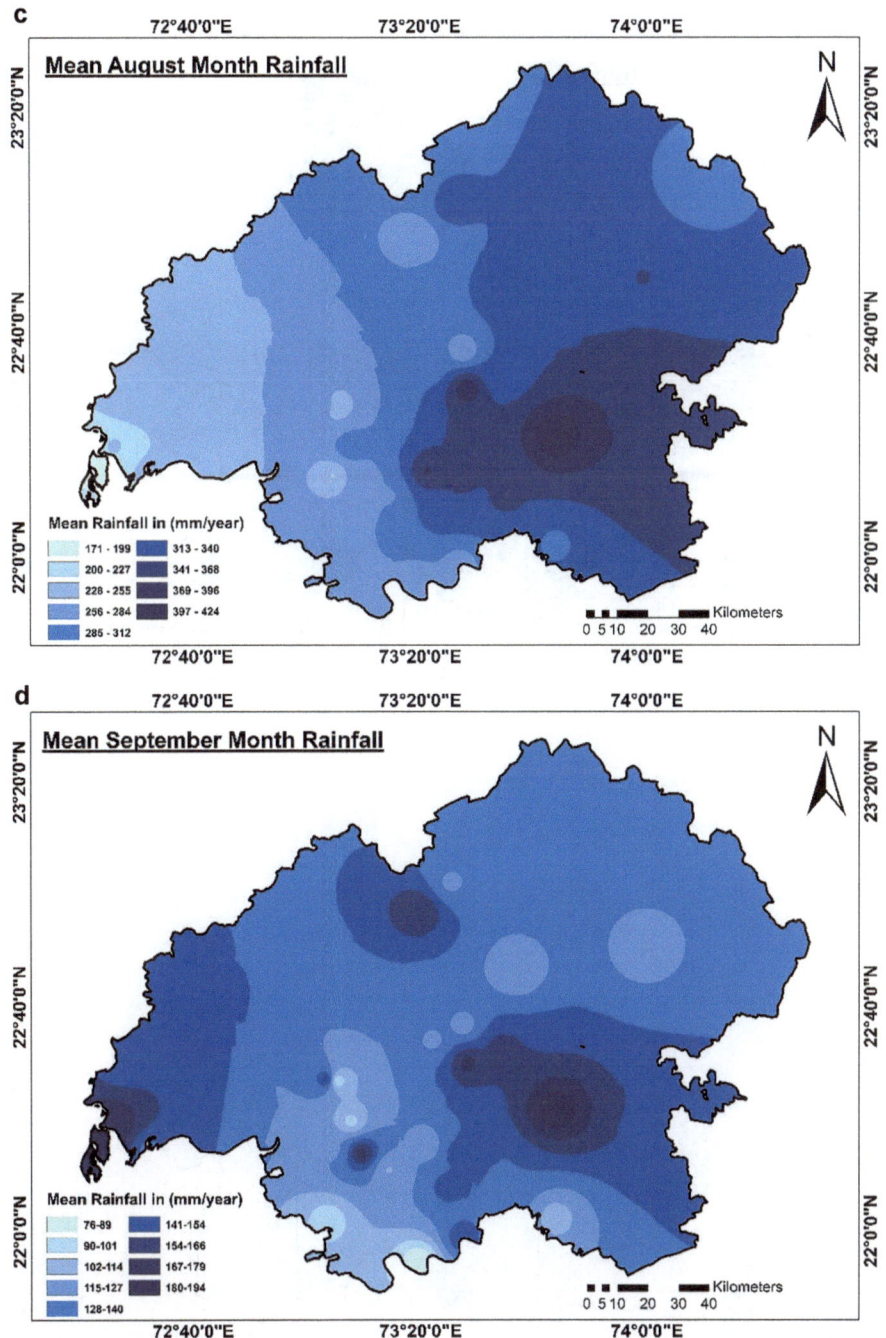

Fig. 5.4 (continued)

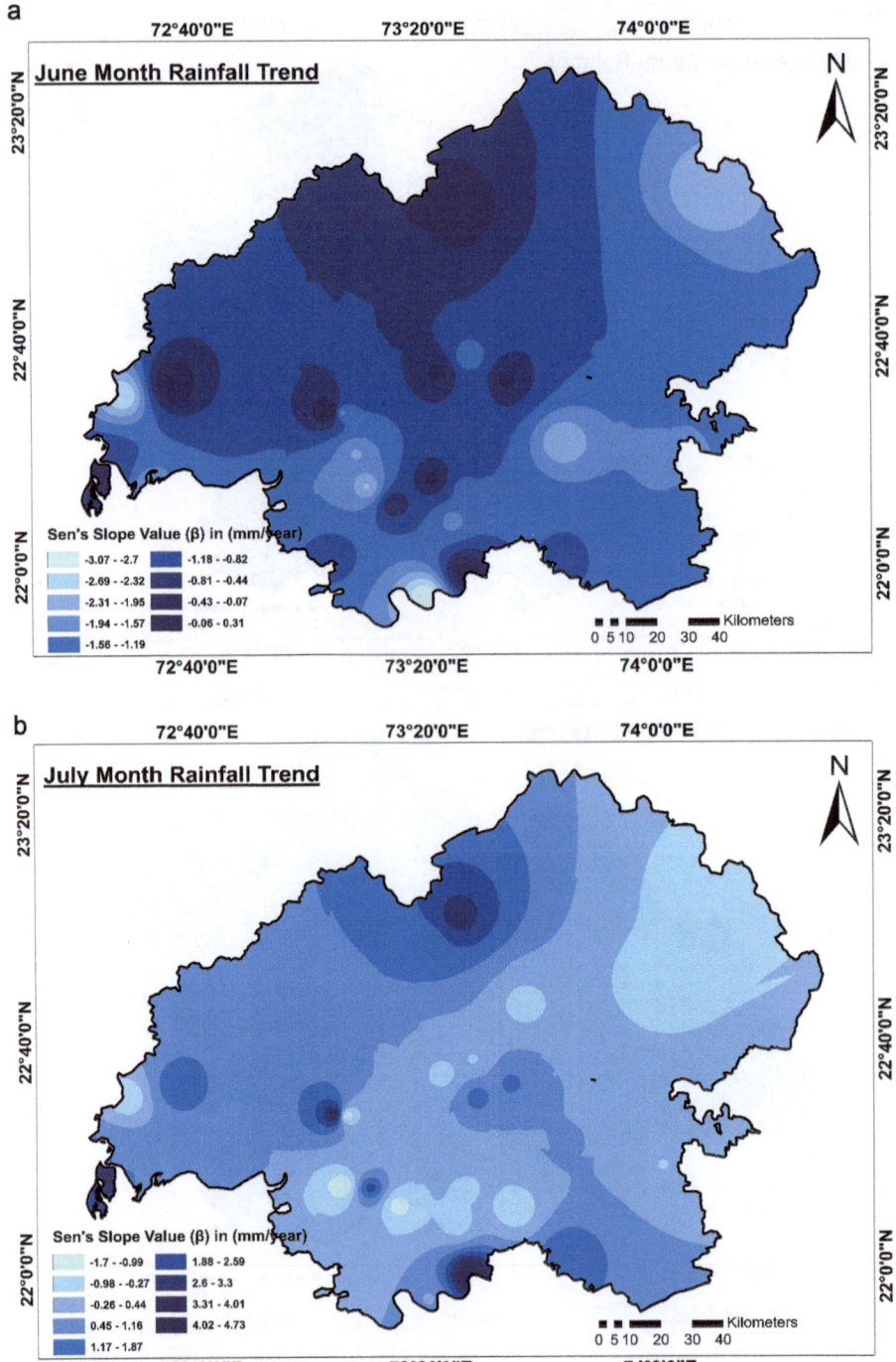

Fig. 5.5 (**a**) Spatial distribution maps of mean monthly rainfall trend in June. (**b**) Spatial distribution maps of mean monthly rainfall trend in July. (**c**) Spatial distribution maps of mean monthly rainfall trend in August. (**d**) Spatial distribution maps of mean monthly rainfall trend in September

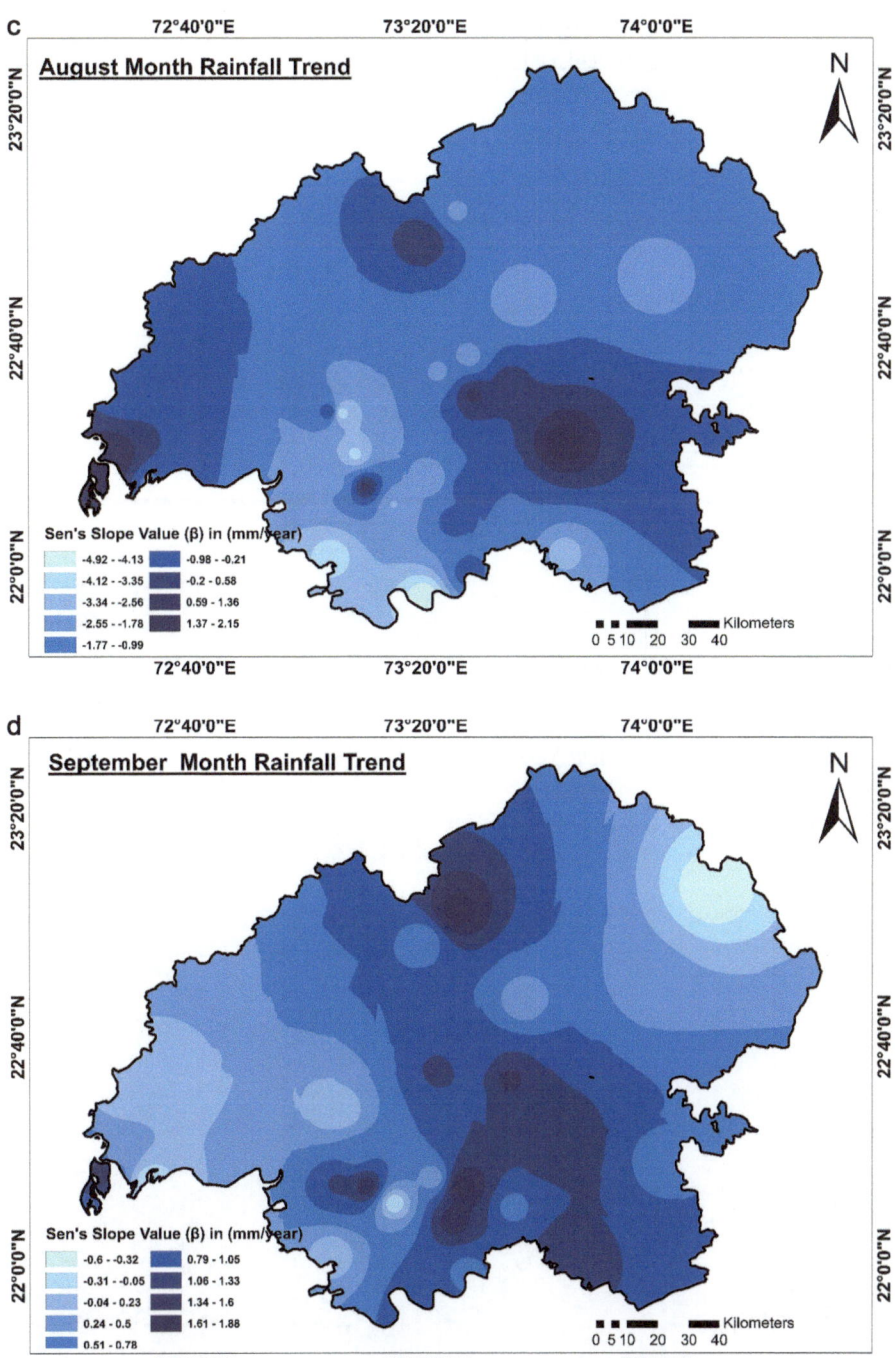

Fig. 5.5 (continued)

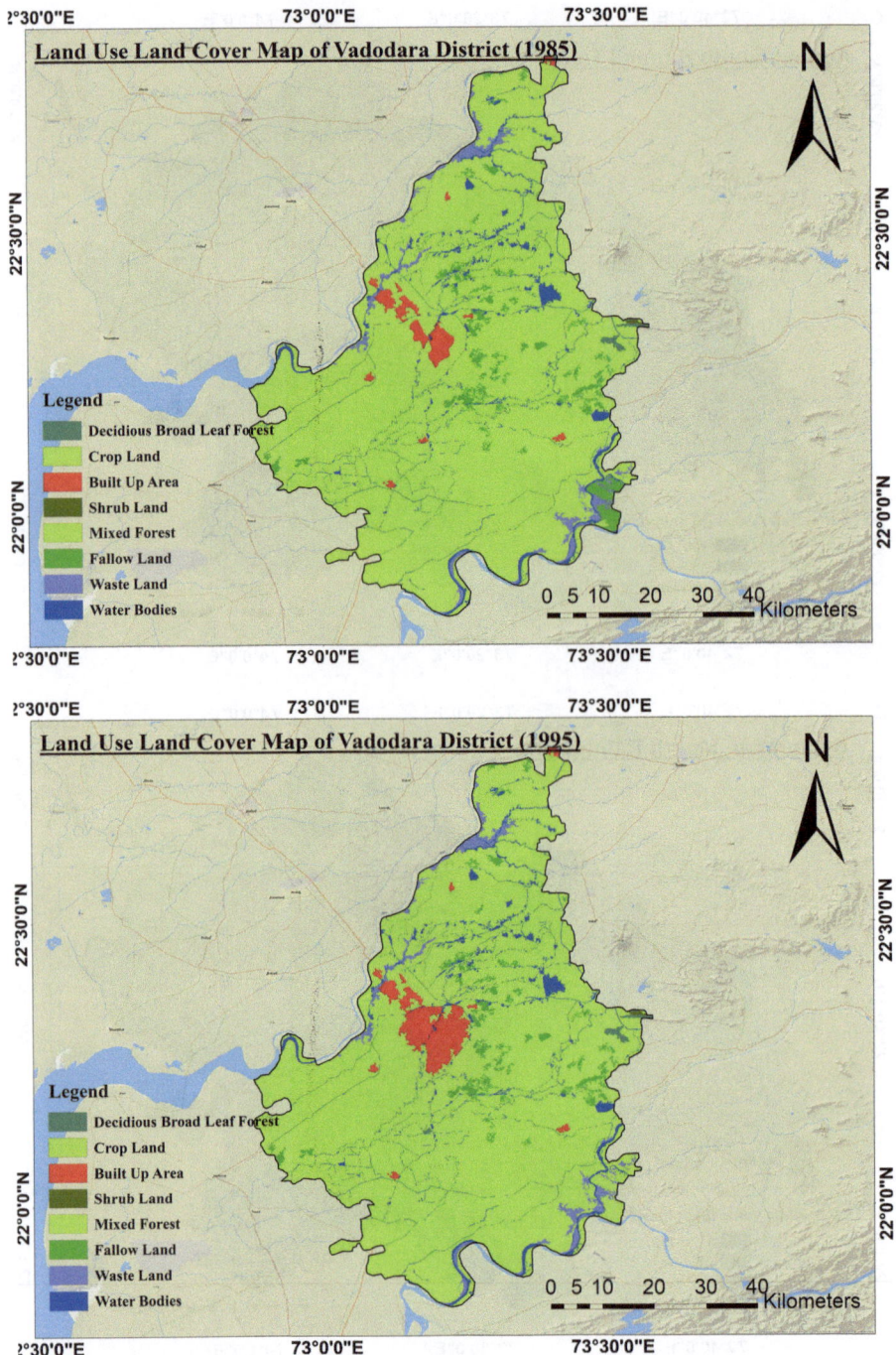

Fig. 5.6 Land use/land cover map of the vadodara district

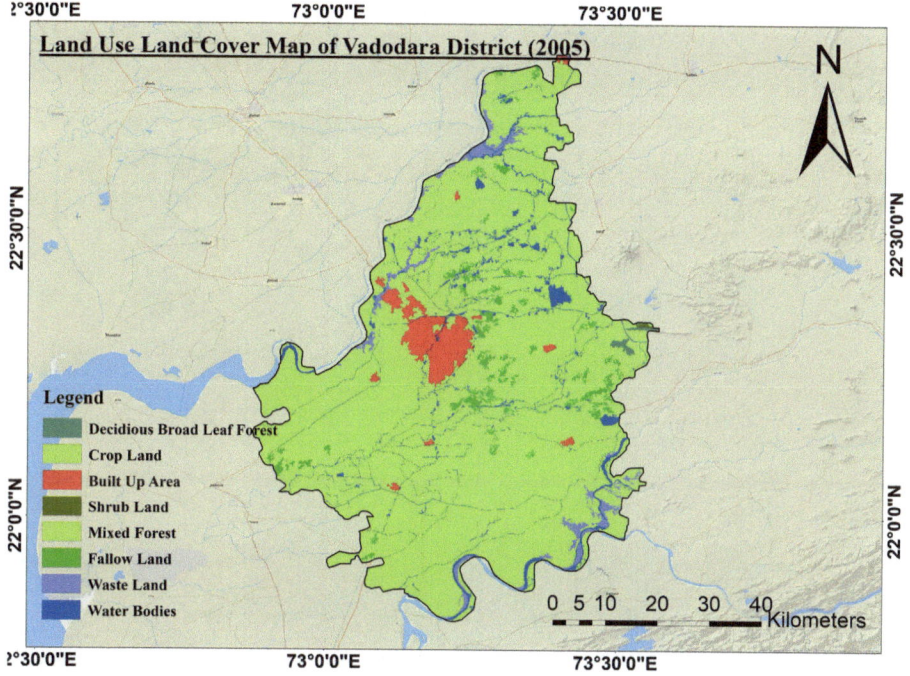

Fig. 5.6 (continued)

Table 5.3 Percentage and percentage change of land use/land cover

Sr. No	Type of land use/land cover (LU/ LC)	Percentage area			Percentage change between 1985 and 1995	Percentage change between 1995 and 2005	Percentage change between 1985 and 2005
		1985	1995	2005			
1	Deciduous broadleaf forest	0.2361	0.2672	0.2378	13.1823	−11.0100	0.7209
2	Crop land	87.3968	86.4496	86.2507	−1.0838	−0.2301	−1.3114
3	Built-up land	1.6484	3.2684	3.3593	98.2743	2.7821	103.7906
4	Mixed forests	0.0425	0.0425	0.0435	0.0000	2.2857	2.2857
5	Shrubland	1.0506	1.1143	1.1483	6.0634	3.0548	9.3034
6	Fallow land	3.0810	2.3134	2.4155	−24.9132	4.4141	−21.5988
7	Wasteland	2.3190	2.3190	2.3416	0.0000	0.9750	0.9750
8	Water bodies	4.2407	4.2407	4.2183	0.0000	−0.5275	−0.5275

Chapter 6
Results and Discussion

In this research work, the trend analysis is carried out for 30 rain gauge stations located in the semiarid region of Gujarat. The daily rainfall data is collected from the State Water Data Centre (SWDC) and Indian Meteorological Department (IMD) for 21 and 9 stations, respectively.

The Mann-Kendall test results for SW monsoon rainfall data as per Table 8 show that 90% of stations (27) have no significant trend ($p = 0.05$). Only three stations have a significant decreasing trend, Dakor from Kheda district ($Z = -2.76$, $\beta = -9.03$), Zalod from Dahod district ($Z = -2.06$, $\beta = -6.7$), and Karjan from Vadodara district ($Z = -1.99$, $\beta = -7.43$).

The Mann-Kendall test in Vadodara district shows a significant decreasing trend in five out of nine stations in the month of June.

In June, the highest magnitude of increasing trend is observed at Vasad station with Z statistic and Sen's slope values of 0.49 and 0.33 mm/year, respectively. The highest magnitude of decreasing trend is observed at Mehemdabad and Sinor stations with Z statistic and Sen's slope values of -3.76 and -3.075 mm/year, respectively.

In July, the highest magnitude of increasing trend is observed at Chandod station with Z statistic and Sen's slope values of 2.11 and 4.73 mm/year, respectively. The highest magnitude of decreasing trend is observed at Padra station with Z statistic and Sen's slope values of -0.99 and -1.72 mm/year, respectively.

In August, the highest magnitude of increasing trend is obtained at Kanewal station with Z statistic and Sen's slope values of 1.58 and 2 mm/year, respectively. The highest magnitude of decreasing trend is obtained at Sinor station with Z statistic and Sen's slope values of -2.56 and -4.92 mm/year, respectively.

In September, the highest magnitude of increasing trend is observed at Kanewal and Karad dam stations with Z statistic and Sen's slope values of 1.6 and 1.88 mm/year, respectively. The highest magnitude of decreasing trend is obtained at Zalod

K. Naveena et al., *Spatio-temporal Trend Analysis of Rainfall using R Software and ArcGIS*, SpringerBriefs in Climate Studies, https://doi.org/10.1007/978-3-031-48259-5_6

station with Z statistic and Sen's slope values of −0.67 and −0.6 mm/year, respectively.

From the rainfall trend analysis, it is observed that two out of nine stations in Vadodara district show a significant decreasing trend in SW monsoon, whereas in June, five out of nine stations show a significant decreasing trend. The change point results also indicate that Sinor, Savli, and Karjan stations of Vadodara district have a significant change point and decreasing trend. The land use/land cover maps are prepared at decadal intervals of 1985, 1995, and 2005 for the Vadodara district to detect the possible changes in the LU/LC patterns.

Chapter 7
Conclusion

Monsoon and monthly rainfall trend analysis has been carried out using data of 30 rain gauge stations for the semiarid region of Gujarat. Cluster analysis is performed to analyze the variability of the mean rainfall. The stations have been divided into 2 clusters with 17 and 13 stations in each cluster, which significantly differ from each other. The first cluster shows a higher mean rainfall than the second cluster.

The result of the monsoon rainfall trend analysis shows that the rainfall decreases at an average rate of −1.79 mm/year. It is observed that 21 stations show a decreasing trend in monsoon, out of which 4 stations (Dakor, Zalod, Karjan, and Sinor) show a significant decreasing trend.

The results of the monthly rainfall trend analysis as per the Mann-Kendall test show that in June the rainfall decreases at an average rate of −1.23 mm/year, where 27 stations show a decreasing trend, out of which 10 stations show a significant decreasing trend. In July, the rainfall increases at an average rate of 0.39 mm/year, where 15 stations show an increasing trend, out of which 2 stations show a significant increasing trend. In August, the rainfall decreases at an average rate of −1.24 mm/year, where 24 stations show a decreasing trend with 2 stations showing a significant decreasing trend. In the month of September, the rainfall increases at an average rate of 0.69 mm/year, where 24 stations show an increasing trend.

The spatial distribution maps of mean rainfall show the highest rainfall near the southern region (Chhota Udepur district), while lower rainfall can be observed in the northern and eastern regions (Anand district).

From the trend analysis results, it can be concluded that there is an overall decrease in the monsoon rainfall over the semiarid region, In the month of June, an appreciable decrease in rainfall is observed with ten stations showing a significantly decreasing trend. In the Vadodara district, 5 out of 9 stations show a significantly decreasing trend in June, 2 stations show a significantly decreasing trend in August, and 2 stations show a significantly decreasing trend in monsoon. Land use/land cover maps at decadal intervals of 1985, 1995, and 2005 for the Vadodara district

© The Author(s), under exclusive license to Springer Nature
Switzerland AG 2023
K. Naveena et al., *Spatio-temporal Trend Analysis of Rainfall using R Software and ArcGIS*, SpringerBriefs in Climate Studies,
https://doi.org/10.1007/978-3-031-48259-5_7

are prepared to link the changes in the rainfall. From the land use/land cover maps, it is observed that the built-up area increased by 103% between 1985 and 2005, which can be a reason for the significant changes occurring in the Vadodara district. The research work was done with the hope that it will be helpful for a better water resource management.

References

Bandyopadhyay N, Bhuiya C, Saha AK (2016) Heat waves, temperature extremes and their impacts on monsoon rainfall and meteorological drought in Gujarat, India. Nat Hazards 82:367–388. https://doi.org/10.1007/s11069-016-2205-4

Bhargava N, Bhargava R, Tanwar PS, Narooka (2016) Comparative study of inverse power of IDW interpolation method in inherent error analysis of aspect variable, vol 2. https://doi.org/10.1007/978-981-10-3920-1

Biswas RN, Islam N, Mia J (2020) Modeling on the spatial vulnerability of lightning disaster in Bangladesh using GIS and IDW techniques. Spat Inf Res 28:507–521. https://doi.org/10.1007/s41324-019-00311-y

Dhendra M, Sunarna H, Ekaprana W, Muljono (2018) The determination of cluster number at k-mean using elbow method and purity evaluation on headline news. International Seminar on Application for Technology of Information and Communication

Emrah H, Dervis K (2016) A comprehensive survey of traditional, merge-split and evolutionary approaches proposed for determination of cluster number. Swarm Evolution Computat. https://doi.org/10.1016/j.swevo.2016.06.004

Gajbhiye S, Meshram C, Mirabbasi R (2016) Trend analysis of rainfall time series for Sindh river basin in India. Theor Appl Climatol 125:593–608. https://doi.org/10.1007/s00704-015-1529-4

Gocic M, Trajkovic S (2012) Analysis of changes in meteorological variables using Mann-Kendall and Sen's slope estimator statistical tests in Serbia. Glob Planet Chang 100:172–182. https://doi.org/10.1016/j.gloplacha.2012.10.014

Hamed KH (2009) Enhancing the effectiveness of prewhitening in trend analysis of hydrologic data. J Hydrol 368:143–155. https://doi.org/10.1016/j.jhydrol.2009.01.040

Hamed KH, Ramachandra Rao A (1998) A modified Mann-Kendall trend test for autocorrelated data. J Hydrol 204(1–4):182–196. https://doi.org/10.1016/S0022-1694(97)00125-X

Hao W, Hui Q (2016) Innovative trend analysis of annual and seasonal rainfall and extreme values in Shaanxi, China, since the 1950s. Int J Climatol. https://doi.org/10.1002/joc.4866

Hassan M, Noreen Z, Rashid A (2020) Regional frequency analysis of annual daily rainfall maxima in Skane. Sweden Inter J Climatol:1–14. https://doi.org/10.1002/joc.7074

Jain SK, Kumar V, Saharia M (2012) Analysis of rainfall and temperature trend in Northeast India. Int J Climatol. https://doi.org/10.1002/joc.3483

Jeneiova K, Kohnova S, Miroslav Sabo Detecting (2014) Trends in the annual maximum discharges in the Vah River basin. Slovakia Acta Silvatica et Lignaria Hungarica:133–144. https://doi.org/10.2478/aslh-2014-0010

© The Author(s), under exclusive license to Springer Nature Switzerland AG 2023
K. Naveena et al., *Spatio-temporal Trend Analysis of Rainfall using R Software and ArcGIS*, SpringerBriefs in Climate Studies,
https://doi.org/10.1007/978-3-031-48259-5

Kannan S, Ghosh S (2011) Prediction of daily rainfall state in a river basin using statistical down-scaling from GCM output. Stoch Environ Res Risk Assessment 25:457–474. https://doi.org/10.1007/s00477-010-0415-y

Kendall (1975) Rank correlation methods charles griffin, London, pp 34–37

Koyel S, Lunagari M (2020) Association between drought and agricultural productivity using remote sensing data: a case study of Gujarat state of India. J Water Climate Change 11(S1):189–202. https://doi.org/10.2166/wcc.2020.157

Li J (2017) Clustering and forecasting for rain attenuation time series data, Stockholm, Sweden, p 11. https://doi.org/10.1016/j.jeconom.2020.06.008

Majed A, Kumari M, Mallic M, Ramakrishnan R, Islam S, Kumar C (2021) Time series trend analysis of rainfall in last five decades and its quantification in Aseer region of Saudi Arabia. Arab J Geosci 14:519. https://doi.org/10.1007/s12517-021-06935-5

Mann HB (1945) Nonparametric tests against trend. Econometrica 13:245–259. https://doi.org/10.2307/1907187

Mehta L, Srivastava S, Adam HN, Alankar SB, Ghosh U, Kumar VV (2019) Climate change and uncertainty from 'above' and 'below': perspectives from India. J Regional Environ Change 19:1533–1547

Patel PS, Rana SC, Josh GS (2021) Temporal and spatial trend analysis of rainfall on Bhogavo River watersheds in Sabarmati lower basin of Gujarat, India. Acta Geophys 69:353–364. https://doi.org/10.1007/s11600-020-00520-2

Saha S, Chakraborty D, Paul RK et al (2018) Disparity in rainfall trend and patterns among different regions: analysis of 158 years' time series of rainfall dataset across India. Theor Appl Climatol 134:381–395. https://doi.org/10.1007/s00704-017-2280-9

Sen Z (2012) Innovative trend analysis. J Hydrol Eng 17(9):1042–1046. https://doi.org/10.1061/(ASCE)HE.1943-5584.0000556

Shi C, Wei B, Wei S et al (2021) A quantitative discriminant method of elbow point for the optimal number of clusters in clustering algorithm. Wireless Com Network 31. https://doi.org/10.1186/s13638-021-01910-w

Shin M-J, Joseph HA, Guillaume FW, Croke AJ, Jakema (2013) Addressing ten questions about conceptual rainfall–runoff models with global sensitivity analyses in R. J Hydrol 503:135–152. https://doi.org/10.1016/j.jhydrol.2013.08.047

Suryavanshi S, Pandey A, Chaube UC et al (2014) Long-term historic changes in climatic variables of Betwa Basin, India. Theor Appl Climatol 117:403–418. https://doi.org/10.1007/s00704-013-1013-y

Tasiya RF, Naveena K, Rana SC (2023a) Rainfall trend analysis in Gujarat's semi-arid zone: a modified approach with auto-correlation consideration. Inter J Hydrol Sci Technol. https://doi.org/10.1504/IJHST.2023.10056997

Tasiya RF, Rana SC, Naveena K (2023b) Change point and trend analysis of rainfall for the Semi-Arid zone of Gujarat state. Water Energy Inter 65(11):6–14

Vaidya VB, Suvarn D, Kulshreshtha MS (2012) Evaluation of frequency analysis of distinctive rainfall intensity for various stations of Gujarat. IJISET - Inter J Innovat Sci Eng Technol 7(12):2348–7968

Yue S, Wang CY (2002) Applicability of prewhitening to eliminate the influence of serial correlation on the Mann-Kendall test. Water Resour Res 38(6):41–47. https://doi.org/10.1029/2001WR000861

Zinabu A, Dioha MO (2020) Climate change and trend analysis of temperature: the case of Addis Ababa, Ethiopia. Environ Syst Res 9(27). https://doi.org/10.21203/rs.3.rs-41363/v1